Advances in African Economic, Social and Political Development

Africa is emerging as a rapidly growing region, still facing major challenges, but with a potential for significant progress – a transformation that necessitates vigorous efforts in research and policy thinking. This book series focuses on three intricately related key aspects of modern-day Africa: economic, social and political development. Making use of recent theoretical and empirical advances, the series aims to provide fresh answers to Africa's development challenges. All the socio-political dimensions of today's Africa are incorporated as they unfold and new policy options are presented. The series aims to provide a broad and interactive forum of science at work for policymaking and to bring together African and international researchers and experts. The series welcomes monographs and contributed volumes for an academic and professional audience, as well as tightly edited conference proceedings.Relevant topics include, but are not limited to, economic policy and trade, regional integration, labor market policies, demographic development, social issues, political economy and political systems, and environmental and energy issues.

More information about this series at http://www.springer.com/series/11885

Almas Heshmati
Editor

Economic Integration, Currency Union, and Sustainable and Inclusive Growth in East Africa

Springer

Editor
Almas Heshmati
Department of Economics
Jönköping International Business School
Jönköping, Sweden

ISSN 2198-7262 ISSN 2198-7270 (electronic)
Advances in African Economic, Social and Political Development
ISBN 978-3-319-30431-1 ISBN 978-3-319-30432-8 (eBook)
DOI 10.1007/978-3-319-30432-8

Library of Congress Control Number: 2016939427

Printed on acid-free paper

This Springer imprint is published by Springer Nature
The registered company is Springer International Publishing AG Switzerland

Acknowledgements

This edited volume is an outcome of a project funded by the Swedish International Development Cooperation Agency (SIDA). The project involves development of Doctoral Programs in Economics and Management, Research Capacity Building and Training and Career Development Initiative at the University of Rwanda, College of Business and Economics (UR-CBE). The project is implemented jointly by Jönköping University, Jönköping International Business School (JIBS) and the College of Business and Economics at the University of Rwanda. The Doctoral Programs aim at building local teaching and supervision capacities, while the Career Development Initiative aims at assisting faculty and staff of UR-CBE to develop their research and academic writing skills.

The chapters in this volume were selected from among papers presented at the conference 'Recent Trends in Economic Development, Finance and Management Research in Eastern Africa', in Kigali, Rwanda, on 4–6 May 2015. The conference was organized jointly by JIBS and CBE. The submissions focused on recent trends in economic development in the developing economies of East Africa. This was the first conference in what is planned to be a yearly event. Theoretical, methodological and empirical research and policy or practice-oriented papers were invited, provided they were based on sound conceptual foundations with well-thought methods. Applied and practice-oriented manuscripts could focus on Eastern Africa as a whole or a group of countries or individual economies in the region. Priority was given to studies on Eastern Africa, but submissions from the rest of Africa were also welcome. Papers from regions other than Africa were also welcome, provided they discussed the implications of the research findings in the African context.

Almas Heshmati

Abbreviations

ADF	Augmented Dickey Fuller
AfDB	African Development Bank
ANOVA	Analysis of variance
ANS	Adjusted net savings
AR	Autoregressive
ASEAN	Association of Southeast Asian Nations
CEMAC	Economic and Monetary Community of Central Africa
CET	Common external tariff
COLS	Corrected ordinary least squares
COMESA	Common Market for Eastern and Southern Africa
CTN	Common tariff nomenclature
DOTS	Directory of trade statistics
DRC	Democratic Republic of Congo
EABEW	East African business and economic watch
EAC	East African community
EAMU	East African Monetary Union
EC	European Community
ECCAS	Economic Community of Central African States
ECOMOG	Economic Community of West African States Monitoring Group
ECOWAS	Economic Community of West African States
EEC	European Economic Community
EF	Ecological footprint
EITI	Extractive industries transparency initiative
EMU	European monetary union
EIA	Environmental impact assessment
EU	European Union
FDI	Foreign direct investment
FTA	Free trade agreement
GCC	Gulf Cooperation Council
GDP	Gross domestic product
GEMI	Global environmental management initiative

GHS	General household survey
GLS	Generalized least squares
GOR	Government of Rwanda
GNI	Gross national income
GNS	Gross national savings
HDI	Human development index
HDR	Human development report
HHI	Herfindahl–Hirschman index
HI	Herfindahl index
HIC	High-income countries
HIV/AIDS	Human immunodeficiency virus infection/acquired immune deficiency syndrome
ICRG	International Country Risk Guide
ICT	Information and communication technology
IGAD	Intergovernmental Authority on Development
IKS	Indigenous knowledge system
IMF	International Monetary Fund
LDC	Less developing countries
LIC	Low-income countries
LSDV	Least squares dummy variable
M3	Broad money
MA	Moving average
MIC	Middle-income countries
MDG	Millennium development goal
MU	Monetary union
NAFTA	North American Free Trade Agreement
NNI	Net national income
OAS	Organization of American States
OCA	Optimum currency area
OECD	Organisation for Economic Co-operation and Development
OLS	Ordinary least squares
OPEC	Organization of Petroleum Exporting Countries
PRSP	Poverty reduction strategy paper
RCA	Revealed comparative advantage
SACU	Southern African Customs Union
SADC	South African Development Community
SAP	Structural adjustment programs
SME	Small and medium enterprises
SSA	Sub-Saharan Africa
TFP	Total factor productivity
TIFA	Trade and Investment Framework
UN	United Nations
UNCTAD	United Nations Conference on Trade and Development
UNDP	United Nations Development Program

UNESCO	United Nations Educational Scientific and Cultural Organization
USA	United States of America
VAR	Vector autoregressive
VAT	Value added tax
VECM	Vector error correction model
WAEMU	West African Economic and Monetary union
WAMZ	West African Monetary Zone
WDI	World Development Indicators
WHO	World Health Organization
WTO	World Trade Organization
WVS	World Value Surveys

Contents

List of Contributors

Ahurra Hope Ayebale Faculty of Education, Kyambogo University, Kampala, Uganda

Dan Ayebale Faculty of Business and Administration, Uganda Christian University, Mukono, Uganda

Thomas Bwire Bank of Uganda, Kampala, Uganda

Hector Carcel University of Navarra, ICS, Navarra Center for International Development, Pamplona, Spain

Ewa Cieślik Poznan University of Economics, Poznan, Poland

Chika Ezeanya College of Business and Economics, University of Rwanda, Kigali, Rwanda

Luis A. Gil-Alana Faculty of Economics, University of Navarra, Pamplona, Spain

Jaromír Harmáček Department of Development Studies, Palacky University, Olomouc, Czech Republic

Almas Heshmati Jönköping International Business School, Jönköping, Sweden

Godfrey Madigu University of Navarra, ICS, Navarra Center for International Development, Pamplona, Spain

Pascal Munyankindi National Bank of Rwanda, Kigali, Rwanda

Syed Mansoob Murshed Institute of Social Studies (ISS) of the Erasmus University Rotterdam (EUR), The Hague, The Netherlands

Herman Musahara Organization for Social Sciences Research in Eastern and Southern Africa (OSSRESA), Addis Ababa, Ethiopia

Emmanuel Mushimiyimana University of Rwanda, School of Social, Political and Admnistrative Sciences, Butare, Rwanda

Esther Nafunka Faculty of Business and Administration, Uganda Christian University, Mukono, Uganda

Miroslav Syrovátka Department of Development Studies, Palacky University, Olomouc, Czech Republic

Caleb Tamwesigire University of Rwanda, College of Business and Economics, Kigali, Rwanda

Yvonne Umulisa Jönköping International Business School, Jönköping, Sweden

Introduction and Summary

Almas Heshmati

Abstract This edited volume is a collection of selected studies on economic integration, currency union and sustainable and inclusive growth in East Africa. The volume consists of 11 inter-related studies. These are largely grouped into economic integration and its trade effects; financial sector development and a common currency; research, innovation and knowledge, and the shadow economy; inclusive and sustainable growth; and the conflict-growth nexus and reconstruction. The studies together provide a comprehensive picture of the state of economic development, growth and integration in East Africa. Several studies cover major parts or the entire continent, but the main focus is on economic development and cooperation among countries in East Africa—Burundi, Ethiopia, Kenya, Rwanda and Tanzania. By employing diverse up-to-date data and methods this volume provides a wealth of empirical evidence and provides sound recommendations to researchers and policymakers in East Africa for designing and implementing effective and inclusive strategies and policies for promoting development and economic integration.

Keywords Economic integration • East Africa • Currency union • Common currency • Sustainable development • Inclusive growth • Trade effects • Financial sector development • Innovation and knowledge • Shadow economy • Conflict-growth nexus • Reconstruction

1 Introduction

The core argument of this edited volume is that East Africa is developing rapidly. Economic integration, currency union and sustainability of growth are among the crucial factors in the development of the region. These issues are not well researched. This book with contributions from professionals in the field sheds light on various determinants of economic integration, currency union and

A. Heshmati (✉)
Jönköping International Business School (JIBS), Jönköping University, Jönköping, Sweden

Sogang University, Seoul, Korea
e-mail: almas.heshmati@gmail.com

A. Heshmati (ed.), *Economic Integration, Currency Union, and Sustainable and Inclusive Growth in East Africa*, Advances in African Economic, Social and Political Development, DOI 10.1007/978-3-319-30432-8_1

sustainability and inclusiveness of growth. It fills existing gaps in literature and provides necessary tools for decision makers.

2 Summary of the Contributing Studies

2.1 Part I. Economic Integration and Its Trade Effects

Two studies in the volume discuss economic integration in East Africa. Chapter 2 by Ewa Cieślik, *Consequences of Economic Integration Initiatives in Africa* is an analysis of the closer integration process in the Eastern Africa region. Cieslik did a study on foreign trade and modern indicators and measures to examine similar, concentration and competitive position changes in Intergovernmental Authority on Development (IGAD) against the background of other economic integrations in Africa. Her main goal was to evaluate the economic potential and development possibilities of IGAD countries in terms of foreign trade and trade integration from 1994 to 2013.

Chapter 3, *Currency Union in the East African Community* by Hector Carcel, Luis A. Gil-Alana and Godfrey Madigu examines inflation rates in the five countries that belong to the East African Community (EAC) that is planning to launch a monetary union. Their aim of their study was to examine the persistence in monthly inflation levels. Literature suggests similar inflation patterns as a precondition for a monetary union. This study shows that the countries present a non-mean reversion, confirming that shocks will not recover in the long-run. Moreover, fractional co-integration relationships were found between most country members.

2.2 Part II. Financial Sector Development and Common Currency

This section has two studies. Chapter 4, *Financial Sector Development–The Economic Growth Nexus in Rwanda* by Caleb Tamwesigire, Thomas Bwire and Pascal Munyankindi investigates the link between a number of relevant financial indicators and economic growth in Rwanda over the period 2006–2014. A triangulation of the Vector Error Correction model and the Granger non-causality approaches and impulse response functions were employed to investigate the dynamic interrelationship between various measures of financial sector development and economic growth and for addressing the questions of direction of causality between financial sector development and growth in Rwanda. Empirical results confirm the existence of a nexus between financial sector development and economic growth in the country which warrants a forward looking monetary policy.

In addition to the previously described study by Carcel, Gil-Alana and Madigu, Yvonne Umulisa in Chap. 5, *Effects of a Common Currency on Intra-Regional*

Trade in Africa, presents theoretical arguments and empirical findings suggesting that a monetary union positively effects intra-union trade which in turn leads to tightly correlated business cycles across countries in a union. The gravity model shows the effects of common currency, the free trade agreement and other main determinants of bilateral trade among African countries. The results suggest the need to realize the EAC monetary union.

2.3 Part III. Research, Innovation and Knowledge and the Shadow Economy

Chapter 6, *Research, Innovation and Indigenous Knowledge in Sub-Saharan Africa* is authored by Chika Ezeanya who establishes a relationship between the low level of innovation experienced across Africa and the absence of indigenous knowledge in the education curriculum and in the continent's research and development agenda. The study explores the concept of innovation and examines the experiences of nations with high levels of innovation and establishes that indigenous or home-grown knowledge is foundational for innovation to thrive. The recognition of indigenous knowledge in formal, informal and non-formal education and research in Africa is foundational for creating a generation of Africans who are innovators, inventors and who are self-motivated to conduct research on issues affecting the development of the continent.

In Chap. 7, *The Shadow Economy and Corruption as Development Impediments*, Almas Heshmati discusses shadow economic activities which have been on the rise causing many problems for society and the state and also raising questions about trust in its organizations. This study is a review of recent studies investigating various theoretical and empirical aspects of shadow economic activities, their measurements and development across developed, developing and transition economies. The main focus is on a number of areas related to the shadow economy and their relationship with public performance and economic growth. The review leads to the identification of several indicators with negative or positive associations with the size of the shadow economy and its causal effects.

2.4 Part IV. Inclusive and Sustainable Growth

Three studies are related to inclusive and sustainable growth. Chapter 8, *Antecedents of Environmentally Friendly Manufacturing Practices among SMEs in Africa*, by Dan Ayebale, Esther Nafunka and Ahurra Hope Ayebale maintains that environmental management is increasingly becoming an important topic of discussion in the business world today. Stakeholders and policymakers are both demanding more accountability from companies in relation to their effects on the environment. Putting the environment at the heart of a company's marketing drive has become a popular competitive strategy. This paper addresses these issues in a rarely studied context. Specifically, it documents empirical evidence on the nature of small and medium enterprises (SMEs) adopting environmentally friendly manufacturing practices in a developing-country context where firms have weak resource bases and operate in poor regulatory regimes. The findings have significant implications for business practices and public environmental policy.

Chapter 9, *Are East African Countries Sustainable?*, is authored by Miroslav Syrovátka and Jaromír Harmáček who assess the extent to which the development of five East African countries is sustainable. The study summarizes the concepts and measurements of sustainability and analyses two composite indicators (Adjusted Net Saving and Ecological Footprint) as applied to five East African countries (Burundi, Kenya, Rwanda, Tanzania and Uganda). The results show that only one country is unsustainable in Adjusted Net Saving, while the Ecological Footprint shows that either all the five countries sustainable or all of them are unsustainable depending on the interpretation of bio-capacity. It also discusses the implications and usefulness of the indices.

Chapter 10, *African Emergence, Inclusive and Sustainable Development and the Role of Social Science Research*, by Herman Musahara argues that the economic growth experience in several African countries and in the region constitutes an emergence. However, the emergence of Africa from colonialism showed that political emergence in the form of flag independence was not enough without economic change. Research provides evidence of the new economic emergence of Africa. However, it also provides evidence of rampant poverty, inequalities and 'jobless' growth. Africa not only faces the challenge of inclusive development but also how to make economic growth friendly to the environment so as to sustain the resources required by future generations. Musahara identifies areas where social science research can contribute to inclusive and sustainable development.

2.5 Part V. Conflict-Growth Relationship and Reconstruction

This section has two studies. Chapter 11, *Cost-Benefit Analysis of the Integration of Rwanda in the East African Community*, is authored by Emmanuel Mushimiyimana. Regional integration is a policy that provides both political and economic gains for member states. Rwanda was torn by war and genocide in 1994. This paper finds that Rwanda is building herself through the East African Community (EAC) by applying political and economic reforms convenient to the integration. Politically, the will for political federation is still limited. Economically, Rwanda is facing a trade deficit though there has been an increase in trade relations between Rwanda and EAC in the last 5 years. However, production is improving domestically in terms of food processing and construction materials. Industrial creation compensating for a decline in tariff revenue is necessary.

The final Chap. 12, *The Conflict-Growth Nexus in Sub-Saharan Africa*, by Syed Mansoob Murshed discusses conflict in Africa which has recently been explained by factors related to greed and grievances. These by themselves are insufficient to initiate conflict in the absence of institutional failures or a degenerating social contract. This study demonstrates the *inseparability* between economic policymaking and the political process. With regard to short-term peace building policies, the economic reconstruction that follows conflict needs to be broad-based and pro-poor otherwise the grievances that produced the civil war may re-emerge. External interventions and economic aid in sustaining peace treaties are important in the short-run. Long-run policies for conflict resolution need to be endogenous.

3 Contributions and Policy Implications

The inter-related and complementary chapters in this volume shed light on a number areas that are important for the East Africa region's economic development and cooperation. The main areas of interest include economic integration, trade relations, formations of monetary unions, common currency, exchange rate, innovation, shadow economy, inclusive and sustainable growth and the conflict-growth nexus and reconstruction. The wide range of topics cover the main determinants of the region's rapid development. This book with contributions from professionals with deep insights in their field identifies various key determinants of economic integration and inclusiveness of growth. It fills existing gaps in literature and provides necessary tools for decision makers to design suitable regional integration, development and reform policies.

Economic integration which is an irreversible process of regional transformation impacts competitiveness, trade, production, labour markets and other related aspects of intra- and inter-regional relations. A reduced development gap between members and developed countries and the creation of a stable economic environment are among the priorities which are conditional on the proper management of foreign trade and globalization. Lack of a common policy and fragmented production has lowered investment advantages. However, economic development and integration within the community do not seem to be a solution to the political, religious and ethnic problems which are often compounded by natural disasters. The conditions for launching a monetary union include sharing similar characteristics and features like homogeneity in inflation rates and investment flows in natural resource areas so as to provide macroeconomic and financial stability thereby reducing inflationary persistence in the future.

Financial sector and economic development are causally related. Broad money is used by monetary authorities to stimulate economic activities. Thus, maintaining the credibility of monetary policy will require a forward looking monetary policy framework. The empirical analysis yields evidence of the long-run and positive influence of financial development on economic growth. In general, overall and diversified financial development, reform of institutions, liberalization and maintenance of macroeconomic stability rather than components of it play a major role in a sector's effectiveness. Empirical findings also suggest that overall a monetary union positively affects intra-union trade which in turn leads to tightly correlated business cycles across countries in the union. The need for an East African Community Monetary Union is supported and its on-going process should remain the primary target for partner states. The main determinants of bilateral trade among the member countries include the degree of openness, level of production, common borders and languages and distance. The community should avoid macroeconomic problems characterizing its European counterpart.

An analysis of the research, innovation and indigenous knowledge suggests that the recognition of indigenous knowledge in education and research in Africa is foundational for creating a generation of innovators and inventors who are self-motivated to conduct research on issues affecting the continent's development. Empowering indigenous knowledge as a fundamental aspect of research in Africa will promote creativity and innovation. Ease of access to research materials will lead to innovation and creativity when indigenous knowledge is emphasized.

A factor that is negative to development is shadow economic activity that has been on the rise leading to many problems for society, the state and public trust in its organizations. The shadow economy has a significant share of the overall economy and captures all its activities. A review of recent studies investigating various aspects of activities in the shadow economy across countries shows the relationship between public performance and economic growth. Several indicators with negative or positive associations with the size of the shadow economy and their causal effects are identified.

Environmental management in production and consumption areas is increasingly receiving more attention. The business sector is demanding more accountability in relation to the effects on environment and health. It has a positive influence on competitiveness of firms. Empirical evidence on the nature of SMEs adopting environmentally friendly manufacturing practices in a developing country context is provided where firms have weak resource bases and operate in poor regulatory environments. Additional facets of the topic with significant implications for business practice and environmental public policymaking are uncovered. Resource constrained manufacturing SMEs need governments to address regulatory issues and to make necessary investments to safeguard the environment. Guidelines are offered for governmental action to enhance adoption of environmental friendly manufacturing practices at the firm level. This is in line with embarking on a sustainable development path. Different composite measurements of sustainability are presented and applied to the East African region to assess their sustainability. Inferences are drawn about possible optimal sustainability policies and measures.

The fast economic growth experience in Africa can constitute an emergence. However, the continent faces challenges of poverty, inequality, jobless growth, non-inclusive growth and environmental degradation. The role of social science research is emphasized to ensure inclusive and sustainable development in the Eastern Africa region.

Given that integration opportunities are used while overcoming implementation of the needed reforms, achieving efficiency to boost domestic production and challenges as well as unexpected outcomes such as management of trade deficits and declines in tariff revenues, regional integration provides political and economic gains. Inferences are drawn for improving a number of sectors in manufacturing and mining. Regional integration might ease the burdens of conflict which can be explained by greed and grievances which are strengthened in the absence of institutions, degeneration of social contracts and lack of sustained economic growth or high rate of growth and inequality effects. Various economic and political measures are suggested to manage the conflict-growth nexus.

This edited volume is authored by experts in the field who employ diverse up-to-date data and methods to provide empirical results based on secondary data covering several countries in East Africa. The volume contains a wealth of empirical evidence and provides recommendations for researchers and policymakers to design and implement effective strategies and policies to develop the region. The book is a useful resource for researchers and policymakers in national and regional research institutes, universities and in non-governmental, and governmental organizations involved in promoting the region's economic integration and development. In addition, the book will also appeal to a broader audience interested in regional economic development and integration.

Part I
Economic Integration and Its Trade Effects

Consequences of Economic Integration Initiatives in Africa: Trade in the IGAD Region

Ewa Cieślik

Abstract The process of close integration in Eastern Africa has resulted in changes in trade structures and production processes across borders. This paper presents the transformations taking place in trade exchange in one of the initiatives of the African community: the Intergovernmental Authority on Development (IGAD). This paper offers a multidimensional analysis of the changes in foreign trade structures of IGAD states, for example, trade creation effects, changes in competitiveness, trade concentration or the similarities of IGAD's trade with selected economic integration communities in Sub-Saharan Africa. In order to ensure uniformity in this analysis, the study is based on data compiled by international organizations, mainly the United Nations Conference on Trade and Development and the World Trade Organization (WTO).

Keywords IGAD • Foreign trade • Trade creation effect • Trade diversion effect • East Africa

1 Introduction

The last decades of the twentieth century brought a significant acceleration in the processes of globalization. This was exemplified in an increase in regional interactions, especially with regard to bilateral and multilateral trade relations. The political factor ceased to be the main driving force behind the integration of countries and it was replaced by economic determinants. International integration and cooperation are perceived by states as giving an opportunity to boost economic development, even though there is no agreement as to the influence of the openness of an economy (trade and capital flows) on the development of a country in academic literature on the subject. In countries other than the most developed ones the liberalization of foreign trade is not necessarily a factor contributing to faster social and economic development.

E. Cieślik (✉)
Poznan University of Economics, Poznan, Poland
e-mail: ewa.cieslik@ue.poznan.pl

A. Heshmati (ed.), *Economic Integration, Currency Union, and Sustainable and Inclusive Growth in East Africa*, Advances in African Economic, Social and Political Development, DOI 10.1007/978-3-319-30432-8_2

In the Eastern African region there are several barriers to regional cooperation such as disparities in the levels of social and economic development, the problem of poverty, institutional and infrastructural weaknesses, epidemics, political instability and not infrequent military conflicts as well. Quite often, one of the reasons for the failure of integration initiatives was attempts to transplant the integration solutions worked out in developed countries to Africa. These models, consistent with literature on theories of international cooperation, did not perform well in the African reality. However, despite these barriers the Intergovernmental Authority on Development (IGAD) was established in 1996[1] and has remained in effect ever since and at present comprises of eight countries—Djibouti, Eritrea, Ethiopia, Kenya, Somalia, Sudan, South Sudan[2] and Uganda. IGAD's mission is to guarantee food security for the people and promote peace. It also oversees security and economic situations (IGAD 2014). IGAD's role, especially in the peace processes in Africa is remarkable (Healy 2013). Among IGAD's many objectives are boosting regional economic cooperation in Eastern Africa, boosting joint development strategies (for example, harmonizing polices with regard to trade and customs) and promoting and realizing the objectives of the Common Market for Eastern and Southern Africa (COMESA) and the African Economic Community (Dundas 2011).

The African continent is often analysed from the angles of poverty, development assistance, problems in the agricultural sector, integration initiatives, political problems, national and international security, trade inequalities and natural resources. Publications which show Africa as an important growth pole and a potential partner in trade and foreign investments have started appearing only recently (Mataen 2012; Cieślik 2014). The first decade of the twenty-first century brought about fundamental qualitative changes in African countries, indicating the huge potential and developmental possibilities emerging out of them. Individual countries in the region recorded the fastest economic growth in the world economy,[3] while most of the developed economies suffered from the economic crisis and its consequences.

On the one hand, African countries have become more important players in international markets while on the other hand, they have turned into a strategic region for the most powerful countries in the world system (Cargill 2010). The next decades, however, are expected to be a period of accelerated economic growth, integration and development in the African continent (Ernst & Young 2013). Therefore, this study is a contribution to further deliberations on the changes in IGAD members' positions in the global economy in terms of foreign trade.

[1] IGAD had its origins in the disastrous droughts that struck the Eastern African region in 1973 and 1984. In 1986 with support from the community the Intergovernmental Authority on Drought and Development was established (IGAD 2014).

[2] The youngest member of this organization, joined IGAD in 2011.

[3] For example, in 2013 Ethiopia's economic growth reached 7 %, Kenya 5.1 %, Djibouti 5 % and Sudan 2.9 %.

The aim of this paper is to analyse foreign trade changes in IGAD member states and the role that the states play in global exports with respect to liberalization processes, their integration initiatives and competitiveness changes. The paper focuses on commodity foreign trade that the investigated states have chosen as the underpinning of their strategies of opening up to the world economy, especially to the African economy. The analysis spans the years 1995–2012 (in some cases, also 1994 and 2013). To make sure that the results are consistent and comparable, data were obtained from databases kept by international organizations, mainly from the United Nations Conference on Trade and Development (UNCTAD) database.

The paper consists of three sections and an introduction and conclusion. First, it presents the IGAD community against the background of other African organizations. This section uses the most important and popular indicators to depict IGAD's position in the African economy. Standard measures as well as more comprehensive methods like concentration and similarities are used in this section. The second section is a study of the main changes in IGAD members' foreign trade, especially in terms of trade creation effects and technology intensity. The next section analyses the changes in the competitiveness position of selected IGAD members' exports (Ethiopia, Kenya and Sudan). The conclusion provides selected recommendations for further regulation in foreign trade in the light of the presented drawbacks.

2 IGAD Integration Initiative Against the Background of Other African Communities

IGAD is one of the recent integration initiatives in the African continent. Unfortunately, there are doubts if this community can be considered an international organization under international law. The doubts are related to its possession of international legal identity status and exercisable power. Generally, it meets the objective criteria[4] and thus it can be said that IGAD has an international legal personality. The *IGAD* Agreement does not spell out any clear rule regarding IGAD's international legal personality. However, Article 3 of the agreement states: 'The Authority shall have the capacity of a legal person to perform any legal act appropriate to its purpose, in accordance with the present Agreement. In particular, it shall have the capacity: (a) to contract; (b) to acquire and dispose of immovable and movable property; and (c) to institute legal proceedings. The Authority shall, in exercising its legal personality, be represented by the Executive Secretary' (Agreement Establishing the IGAD 1996). This provision relates to IGAD's legal personality and capacity in an ambiguous way. Although the IGAD Agreement does not contain express provisions on IGAD's international legal personality, the expressly

[4] IGAD has been created by countries by a treaty under international law and has its own international organs.

granted treaty-making power as well as immunities and the clear recognition of the right to cooperate with other subjects of international law make the possession of international legal personality by IGAD explicit. In addition, in order to execute its objectives, IGAD has adopted relevant institutional acts which comply with the rules of its constitution and applicable international laws (Weldesellassie 2011).

Leaving IGAD's formal aspect aside, it is worth analysing this community in relation to its economic performance. Against the background of other African organizations, the IGAD community seems to be an important economy, though it integrates a relatively small number of states. Especially in terms of regional GDP, its share in the world GDP, real GDP per capita growth rates and FDI flows or populations, IGAD seems to be a relevant community in African society. It also achieves high indicators of intra-regional trade share in comparison to other analysed organizations, indicating close trade interdependence among IGAD members. Certainly, it is inferior to the largest economic blocs in Africa such as the Common Market for Eastern and Southern Africa (COMESA), the Economic Community of West African States (ECOWAS) and the Southern African Development Community (SADC).[5] In terms of the intra-regional trade intensity index, IGAD's intra-regional trade is slightly relatively more important than trade flows with non-member states. Unfortunately, GDP per capita and FDI inflows per capita locate IGAD at a disadvantage in comparison to the other integration initiatives analysed later in the paper (Table 1).

It is worth mentioning that in Africa we observe the 'spaghetti bowl' phenomenon in terms of international agreements and African regional economic organizations (Hartzenberg 2011). Every country from this continent participates in at least two economic communities, 30 states take part in three international agreements, 18 countries in four organizations, with Kenya, the leader in this field, participating in five economic blocs. This accumulation of economic agreements results in obstacles to the trade effect of preferential integration and obfuscates the whole picture of the regional cooperation process (Iddrisu 2012).

The factors that were to become particularly important in boosting the integration processes and socioeconomic development were the growing openness of the economies and their ability to exploit the opportunities of globalization. While analysing the changes in foreign trade channels of IGAD members, it is relevant not only to study the basic indexes of foreign trade exchange, but also to present the processes in which the trade structures of these countries are integrating, becoming similar or drifting apart. In comparison to the other African international organizations, IGAD's openness is limited. It is plausibly an effect of military conflicts within this region (the war between Ethiopia and Eritrea between May 1998 and June 2000, or the ongoing Somali civil war) which coincided with the downturn in the Kenyan economy during the last term of Daniel Toroitich arap Moi's

[5] We should be cautious in analysing intra-regional trade shares because organizations with a higher number of states and larger regions (in terms of total trade) tend to present a higher intra-regional trade share.

Table 1 IGAD region against the background of other African communities (selected indicators 2010)

Indicator	IGAD	CEMAC	COMESA	EAC	ECCAS	ECOWAS	SACU	SADC	WAEMU
Intra-Regional Trade Intensity Index	102.83	178.72	9.70	136.04	193.59	6.66	4.39	13.50	5.81
Intra-regional trade share	8.21	7.05	6.00	11.00	8.85	4.32	3.13	12.15	3.75
Regional GDP in current prices (US$ billion)	160.37	76.15	551.30	79.14	179.20	37.21	396.32	565.04	260.33
Regional GDP per capita	743	1763	1144	580	1216	1027	6859	2174	1090
Regional GDP share in World GDP (%)	0.13	0.06	0.44	0.06	0.14	0.24	0.32	0.45	0.21
Real GDP per capita growth rates (%)	7.22	5.25	5.77	6.29	4.62	6.91	3.38	3.88	4.46
Inward FDI flows (US$ million)	3304	6356	18,003	2578	6162	11,846	2265	8198	1282
Outward FDI flows (US$ million)	n.a.	n.a.	5283	n.a.	n.a.	1288	−73	2591	−4
Inward FDI flows per capita (US$)	15.30	150.60	37.35	18.88	41.80	39.61	39.20	31.54	5.37
Regional population (in thousands)	215,912.1	42,204.3	482,012.1	136,532.4	147,424.3	299,069.2	57,780.4	259,896.0	238,826.15
Regional share in world population (%)	3.14	0.63	7.02	1.99	2.15	4.35	0.84	3.78	3.48

Note: The intra-regional trade intensity index is used to determine whether the value of intra-regional trade is greater or smaller than would be expected on the basis of the region's importance in world trade. Formula of intra-regional trade intensity index:

$$ITII_{i,t} = \frac{\left(\frac{IT_{i,t}}{T_{i,t}}\right)}{\left(\frac{T_{i,t}}{T_{w,t}}\right)}$$

where $IT_{i,t}$ denotes region i's intra-regional trade in year t; $T_{i,t}$ denotes region i's total trade in year t (i's total imports plus total exports); $T_{w,t}$ denotes the world's total trade in year t (world's total imports plus total exports)

The value ranges from 0 to $\frac{T_{w,t}}{T_{i,t}}$. When $ITII_{i,t}$ is equal to zero in the case of no intra-regional trade; when $ITII_{i,t}$ is equal to one (or 100) if the region's weight in its own trade is equal to its weight in world trade (geographic neutrality); when $ITII_{i,t}$ is higher than one (or 100) if intra-regional trade is relatively more important than trade flows with the rest of the world

Source: Author's study on the basis of United Nations (2014), RIKS Platform (2014), The World Bank (2014)

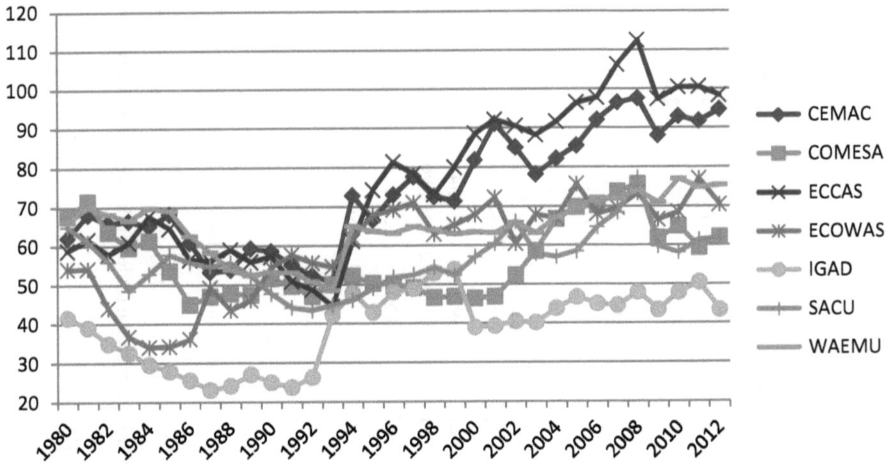

Fig. 1 Openness of selected international economic communities in Africa (1980–2012) (foreign trade as a % of GDP). *Source*: Author's study on the basis of UNCTAD (2014)

presidency. After IGAD was established, we observe a considerable increase in openness or, more precisely, a shift from near autarky to limited trade exchange, but after 1999 the openness has been maintained between 40 and 50 % (Fig. 1).

As a matter of fact, IGAD's share of foreign trade in global trade is not significant and has been subject to considerable fluctuations in the last decade. Members of the organization are more active in terms of imports of goods and services (more than 0.25 % of global imports in 2012) than in terms of exports (about 0.1 % of global exports). IGAD's share of foreign trade in total developing African states' trade, however, points to the important position of the organization in the continent. In 2012, the share of the countries in question in African exports was about 3.2 % and it has been quite stable since 2000. We can observe a rising trend in imports: IGAD states' share of imports in developing African countries' imports increased from 6.5 % in 2000 to 7.5 % in 2012. This increase can be explained primarily by the increasing share in imports of Kenya, Sudan and Uganda (UNCTAD 2014).

COMESA seems to be IGAD's most important export market, which should not be surprising, because this bloc is the largest in the Eastern African region and also covers almost all IGAD countries (except for Somalia). In 2012, IGAD directed to this community primarily food and livestock (32 % of the total exports to the bloc), manufactured goods (21 %) and chemicals (10 %). The South African Development Community (SADC) remained a significant export partner for IGAD, where it exported mainly manufactured goods (25 % of the total exports to the community) and chemical products (15 %). In terms of imports COMESA is also a lead market which provided the IGAD region with machinery and transport equipment (70 % of the total imports from the organization) in 2012. The second significant African bloc that is active in IGAD imports is SADC. This community from Southern

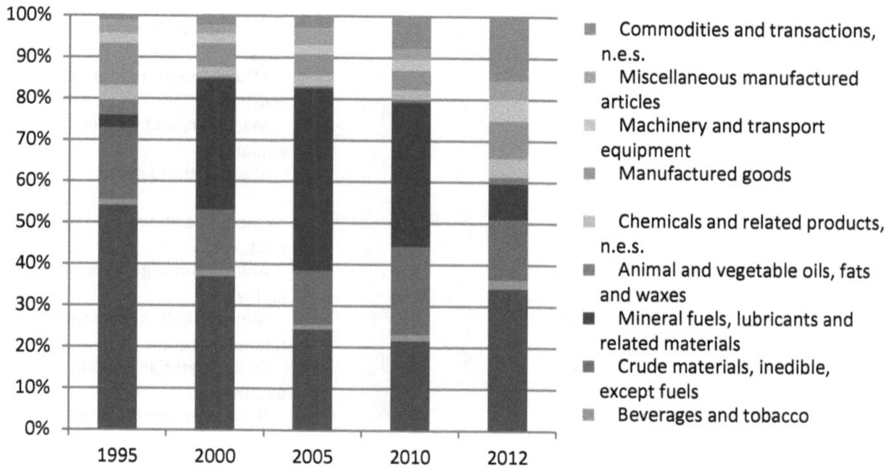

Fig. 2 Export structure of the IGAD countries in 1995–2012. *Source*: Author's study on the basis of UNCTAD (2014)

Africa mainly provided manufactured goods (31 %) and food and live animals (22 %) (UNCTAD 2014).

Analysing the trade structure of IGAD countries, we can observe significant changes between 1995 and 2012. Food remained the predominant export product of the IGAD region; however, its share has decreased through the years. Crude materials (mostly hides, textiles, wood and oil seeds) also have a considerable share in IGAD's exports. The insignificant share of manufacturing goods and still little progress in this field over the past decade indicates that IGAD's exports are not advanced (Fig. 2). The import structure of IGAD states, in turn, has also not changed considerably lately. Machinery and manufactured goods, chemicals and fuels continue to shape the structure of imported commodities. All mentioned products, except for fuel, have not changed their share in imports significantly. Only fuel's share has increased significantly since 1995 (Fig. 3).

The diversity of the revealed comparative advantages is also reflected in the degree of similarities between foreign trade commodity structures of IGAD states. In this study, the foreign trade commodity structures of the largest economic blocs in Africa were adopted as the model. This analysis enabled us to indicate the economic integration blocs similar to IGAD in terms of foreign trade structures (applying Standard International Trade Classification, Rev.3). The Euclidean metric formula was used in the study of the degree of similarity. Comparing 1995 and 2012, it can be observed that the structure of commodities exported by IGAD was almost exact to the East African Community's (EAC) structure, which the IGAD members (Kenya and Uganda) largely shape. We also observed that IGAD is becoming more similar to the Southern African Customs Union (SACU), SADC and West African Economic and Monetary Union (WAEMU) models. IGAD has clearly drifted away from the Economic and Monetary Community of Central

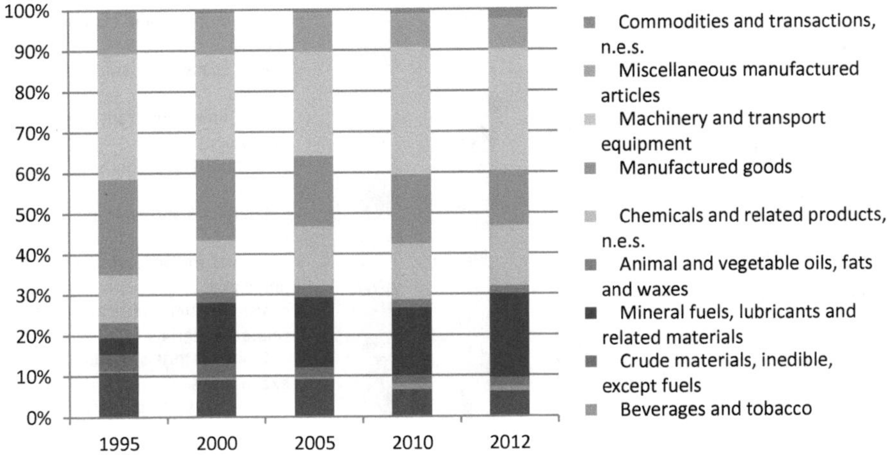

Fig. 3 Import structure of the IGAD countries in 1995–2012. *Source*: Author's study on the basis of UNCTAD (2014)

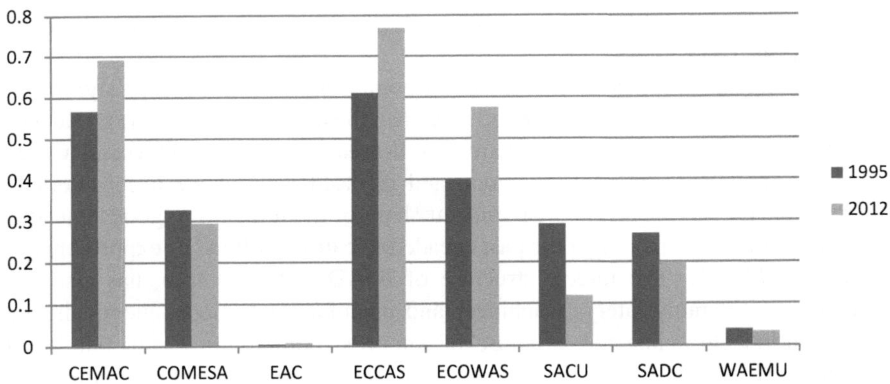

Fig. 4 Indicators of similarity of the export commodity structure of the IGAD states in 1995 and 2012 *Source*: author's own calculations on the basis of UNCTAD (2014) using SITC Rev.3. The Euclidean metric $\sqrt{\sum_{i=1}^{n}(x_i - y_i)^2}$ has been used as the measure of similarity. The closer the value is to one, the more different are the export structures of the analysed blocs. The closer the value is to zero, the more similar the commodity structures

Africa (CEMAC), COMESA, Economic Community of Central African States (ECCAS) and Economic Community of West African States (ECOWAS) models (Fig. 4). COMESA's exports are much more concentrated on fuels and manufactured goods and food and live animals do not hold important positions in the organization's exports. ECOWAS, CEMAC and ECCAS focus generally on mineral fuel exports and other groups of products have a limited share in their export structures (UNCTAD 2014).

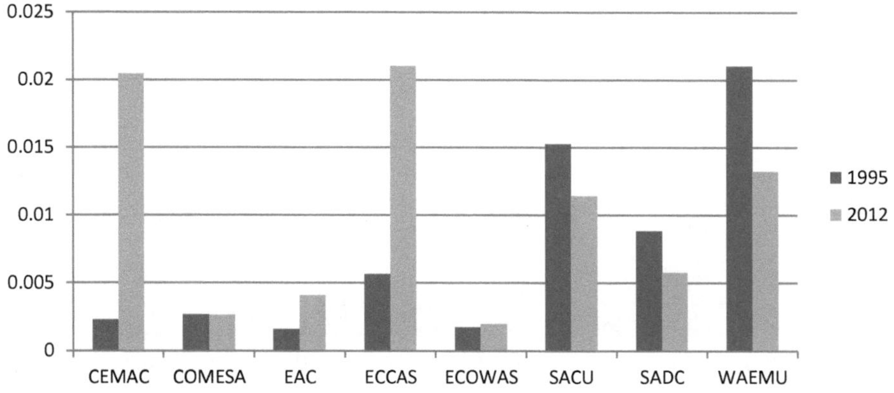

Fig. 5 Indicators of similarity of the import commodity structure of the IGAD states in 1995 and 2012. *Source*: Author's calculations on the basis of UNCTAD (2014) using SITC Rev.3. The Euclidean metric $\sqrt{\sum_{i=1}^{n}(x_i - y_i)^2}$ has been used as the measure of similarity. The closer the value is to one, the more different are the export structures of the analysed blocs. The closer the value is to zero, the more similar the commodity structures

As for the similarity in imports, we can observe small values of the Euclidean metric, which implies a considerable similarity in the import structures of the examined countries with reference to the analysed integration blocs. Among these organizations, IGAD is characterized by the highest similarity with ECOWAS, COMESA and certainly EAC models. Further, in the analysed period, IGAD experienced the greatest decrease in similarity between its import structure and CEMAC and ECCAS models. Generally speaking, the relatively low values of the indicators which describe similarity in import structures mean that the structures of imported commodities deviate insignificantly from the model established by the other African communities (Fig. 5).

In terms of shaping the geographic composition of trade partners, the IGAD region seems to behave similarly to all analysed economic communities in Africa. In exports and imports, Euclidean metric values are small which indicate little distance to the analysed economic African blocs. These indicators also prove that in IGAD, like in all African communities, there is substantial diversification of trade partners. Generally speaking, the analysed organizations are almost identical in terms of the geographic structure of imports and slightly different in terms of export structures (Figs. 6 and 7).

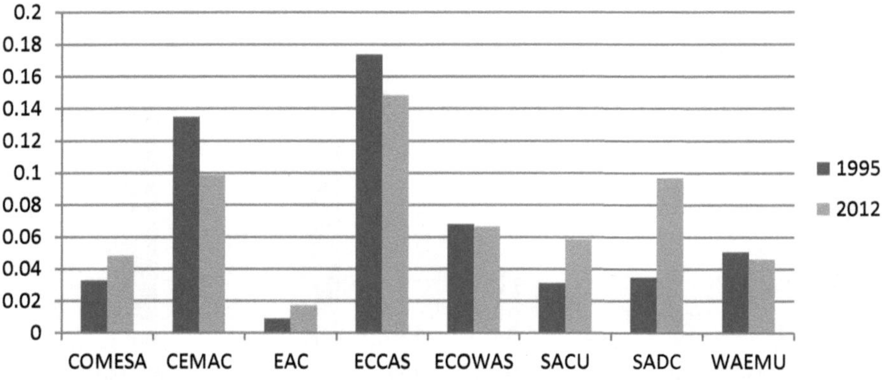

Fig. 6 Indicators of similarity of the export geographical composition of IGAD states (1995 and 2012). *Source*: Author's calculations on the basis of UNCTAD (2014). *Note*: The Euclidean metric $\sqrt{\sum_{i=1}^{n}(x_i - y_i)^2}$ has been used as the measure of similarity. The closer the value is to one, the more different are the export structures of the analysed blocs. The closer the value is to zero, the more similar the commodity structures

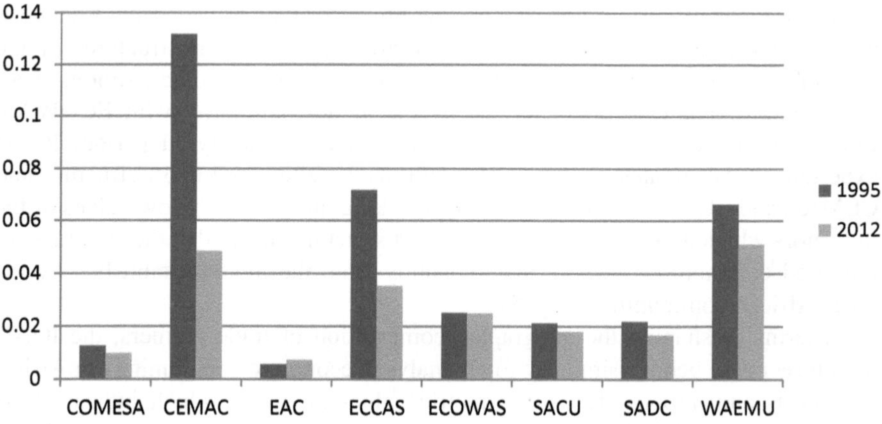

Fig. 7 Indicators of similarity of the import geographical structure of the IGAD states in 1995 and 2012. *Source*: Author's calculations on the basis of UNCTAD (2014). The Euclidean metric $\sqrt{\sum_{i=1}^{n}(x_i - y_i)^2}$ has been used as the measure of similarity. The closer the value is to one, the more different are the export structures of the analysed blocs. The closer the value is to zero, the more similar the commodity structures

3 Foreign Trade Changes in IGAD Member States: Selected Statistics

When analysing IGAD's trade exchange, it is hard not to notice the asymmetry of this exchange. More than 80 % of the entire trade of the community is by Ethiopia, Sudan and Kenya. In 2012, Kenya had a 33.6 % share in exports and imports, Sudan a 25.4 % share and Ethiopia a 22.7 % share (UNCTAD 2014). It would be hard to question the fact that foreign trade within IGAD had no prominent place in the trade of all the members of the organization, which focused on other markets. States in the IGAD region trade little among themselves with the exception of Uganda, Kenya and Ethiopia (Table 2). In 1995, the main trade markets for IGAD were developed countries (70 % of total trade), but by 2012 the share of these markets had decreased to 50 %. The most important export markets for IGAD in 2012 were UEA (14 %), China (8 %) and Tanzania and the Netherlands (both 5 %). In 2012, the IGAD states imported mostly from China (16 %), India (11 %), Saudi Arabia (7 %), the United States (6 %) and the UAE (6 %) (UNCTAD 2014). It should be noted that there has been a growth trend in foreign trade of all IGAD countries from 2000, which was disturbed in 2009 by the global crisis (Chauvin and Geis 2011).

3.1 Trade Diversion and Trade Creation Effect

It is worth answering the question if IGAD's establishment in 1996 resulted in a trade creation effect and a trade diversion effect. Due to the limited length of this paper, only the results of this analysis are presented. This approach relies on the indices of income elasticity of import demand. The trade diversion effect occurs when the average dynamics of IGAD imports from countries outside the organization (e_{c1}—before integration, e_{c2}—after integration) before the establishment of the bloc in relation to the annual average dynamics of IGAD states' GDP before 1996 is higher than these indicators after the establishment of IGAD (E_{c1}—before integration, E_{c2}—after integration). Matrices of intra-regional trades, calculated dynamics of GDP and import from non-members of IGAD for two periods 1994–1996 and 1997–1999 should be created in the analysis. The results are:

Table 2 Percentage share of trade turnover with the IGAD region in the foreign trade of the bloc's members (2012)

Country	Export	Import
Djibouti	11.2 %	3.2 %
Eritrea	0.1 %	1.7 %
Ethiopia	19.4 %	2.7 %
Kenya	15.0 %	1.4 %
Somalia	0.2 %	1.0 %
Sudan	3.4 %	2.3 %
Uganda	20.6 %	13.6 %

Source: Author's calculations on the basis of UNCTAD (2014)

$$\frac{e_{c1}}{E_{c1}} = 0.745353; \frac{e_{c2}}{E_{c2}} = 1.028765$$

$$\frac{e_{c1}}{E_{c1}} < \frac{e_{c2}}{E_{c2}} \rightarrow \text{the trade diversion effect is not observed}$$

This means that the creation of the IGAD region in 1996 did not contribute to the emergence of the trade diversion effect. Intra-trade did not push out trade with states outside the community. This is also proved by the percentage share of imports with IGAD in imports of community members (Table 2). Moreover, we should take into account some important limitations of this analysis. Firstly, the research takes into consideration only short periods of time (3 years before and after the introduction of changes in IGAD's status). For a longer period, perhaps these indicators may be different. Secondly, according to official data, the consensus view is that intra-African trade is generally low. These results, however, are problematic because they do not take into consideration any informal trade flows, which are relatively high in East Africa. For instance, Ugandan informal exports to its neighbors represented US$224 million or 83 % of its total recorded trade to these countries in 2006. In 2009 and 2010, Ugandan informal exports to its neighbors were worth US$790 million and US$520 million respectively (Afrika and Ajumbo 2012).

A popular method of calculating the trade creation effect is gross effect by Balassa (1998). This means that we do not adjudicate whether raising imports has replaced domestic production or third countries' production. First, the total creation effect was calculated. Two years were taken into consideration: 1997 (the first year after the reorganization of IGAD) and 2012 (the last year for which statistics are available for most of the member states). The creation effect for exports and imports was calculated according to the following method assuming linear extrapolation:

(a) *Export creation effect (total)*

$$E_{ex}^{C} = X_{1997}^{c} - [X_{1996}^{c} + \frac{X_{1996}^{c} - X_{1991}^{c}}{5}]$$

where

E_{ex}^{C}—creation effect in total exports

X_{1997}^{c}—real exports of each country to the IGAD region

$X_{1996}^{c} + \frac{X_{1996}^{c} - X_{1991}^{c}}{5}$—the hypothetical exports of each country if the IGAD region does not exist

(b) *Creation effect in i product group export*

$$E_{ex}^{i} = X_{1997}^{i} - [X_{1996}^{i} + \frac{X_{1996}^{i} - X_{1991}^{i}}{5}]$$

where

E_{ex}^{i}—creation effect in *i* product group

X_{1997}^{i}—real exports of *i* product group in 1997

$X^i_{1996} + \frac{X^i_{1996} - X^i_{1991}}{5}$—the hypothetical exports of *i* product group if the IGAD region does not exist

(c) *Import creation effect (total)*

$$E^C_{im} = F^c_{1997} - [F^c_{1996} + \frac{F^c_{1996} - F^c_{1991}}{5}]$$

where

E^C_{im}—creation effect in total imports

F^c_{1996}—real imports of each country from the IGAD region

$F^c_{1996} + \frac{F^c_{1996} - F^c_{1991}}{5}$—the hypothetical imports of each country if the IGAD region does not exist

(d) *Creation effect in i product group import*

$$E^i_{im} = F^i_{1997} - [F^i_{1996} + \frac{F^i_{1996} - F^i_{1991}}{5}]$$

where

E^i_{im}—creation effect in *i* product group

F^i_{1997}—real import of *i* product group in 1997

$F^i_{1996} + \frac{F^i_{1996} - F^i_{1991}}{5}$—the hypothetical imports of *i* product group if the IGAD region does not exist

For 2012 an analogical methodology was applied.

Unfortunately, there are no explicit results of this calculation. Generally, we cannot see any impressive and stable trade creation effects in IGAD members, though some trade liberalization measures have been introduced. The exports creation effect existed at the beginning of the changes in IGAD regulations. This was unlike the imports creation effect, which was stronger in 2012. In 1997, total exports of member states to the IGAD region accounted for US$432 million, (23 % was the creation effect). In 2012, this tendency was the opposite when the total exports of member states to the IGAD region amounted to US$2336 million, which, according to the calculation, was 0.01 % less because of the existence of the IGAD community. It is observed that export creation occurred mostly because of Ethiopia. Only three countries' exports (Ethiopia, Somalia and Sudan) benefited from the changes in IGAD regulations in 1996. In 2012, the situation worsened: only in Somalia and Uganda was there the creation effect while the overall (for all member states together) export creation effect did not occur. In terms of imports, there were opposite effects: the first year after changes in IGAD's status did not bring any import creation, but in 2012 the total effect occurred because of Eritrea, Kenya and Somalia. In 1997 total imports of member states from the IGAD region accounted for US$638.5 million, which, according to the calculation, was 0.02 % less because of the existence of this community. In 2012, total imports of IGAD members from the community amounted to US$1947 million, with the creation effect of 0.05 % (Table 3).

Table 3 The effects of IGAD initiative on trade creation and diversion (1997 and 2012) (US$ in thousands)

Country	Creation effect in total exports		Creation effect in total imports	
	1997	2012	1997	2012
Djibouti	−20,600	−9056	−101	934
Eritrea	−2521	−275	n.a.	50,175
Ethiopia	13,889	−44,150	−41,862	−40,751
Kenya	−43,003	−298,813	15,730	54,806
Somalia	487	1925	−1044	47,497
Sudan	522	−110,399	−1963	−93,053
Uganda	−3602	85,863	−16,831	−159,379
Total	1009	−275	−101	934

Source: Author's calculations on the basis of UNCTAD (2014)

In terms of product groups it is difficult to evaluate which type of goods benefited the most from the IGAD initiative. We also face serious lack of data for many states. The highest creation effect occurred in product group 'crude materials, inedible, except fuels' in 1997 Ethiopia's exports in this product group increased by 24 % because of the creation effect. The largest negative influence of IGAD initiatives on exports was seen in Kenya's manufactured goods' exports in 1997, which were 6.7 % lower than without IGAD's existence. In 2012, Uganda experienced the most significant creation effect in terms of food and live animals—14.2 % of the exports of this product group were the result of the creation effect. Generally, Kenya's exports suffered the most because of IGAD, especially the product groups of 'machinery and transport equipment' and 'food and live animals'. The negative impact on these product groups amounted to 75 and 47 % respectively. Nevertheless, the largest decrease in export values was observed in Sudan in terms of mineral fuels, lubricants and related materials (Table 4).

Research on the import creation effect has some limitations because of the limited availability of statistical data for 1997. This is also the reason why we focus on IGAD's impact on imports only in 2012. The creation effect was visible especially in the more advanced product group of 'machinery and transport equipment' in 2012. This effect was driven mainly by Eritrea's imports of this product group, but almost all IGAD members (except for Uganda) registered import creation effects in this product group. The creation effect in Eritrea's imports of 'machinery and transport equipment' amounted to more than 96 %. The most dramatic decrease in the imports of almost all product groups due to the creation effect was seen in Eritrea and Uganda. Eritrea suffered most from the negative creation effect in the group 'food and live animals', whereas Uganda registered the deepest negative effect in 'manufactured goods' imports (almost 37 % lower than without IGAD's existence). A serious negative creation effect was also seen in Ethiopia's imports of mineral fuels, lubricants and related materials. Generally, 'manufactured goods' was the group of products that suffered the most due to the

Table 4 The exports creation effect in terms of product groups (1997 and 2012) (US$ in thousands)

		Food and live animals	Beverages and tobacco	Crude materials, inedible, except fuels	Mineral fuels, lubricants and related materials	Animal and vegetable oils, fats and waxes	Chemicals and related products, n. e.s.	Manufactured goods	Machinery and transport equipment	Miscellaneous manufactured articles
Djibouti	1997	−2037.9	−182.7	−3.5	n.a.	−842.2	−1159.3	−3182.1	−4414.5	−922.7
	2012	3870.6	34.6	2392.9	−19,603.4	402.9	548.5	1296.2	773.7	1228.1
Eritrea	1997	−104.1	n.a.	−367.1	n.a.	−2273.8	100.4	10.5	913.9	−503.5
	2012	−146.4	50.5	−39.1	n.a.	n.a.	−3.1	−1.8	−20.6	−6.9
Ethiopia	1997	3218.6	7.6	9331.4	1312.1	n.a.	n.a.	11.5	n.a.	3.6
	2012	−62,283.3	−148.2	13,653.7	n.a.	n.a.	−1065.1	1575.6	2443.6	1234.3
Kenya	1997	−2685.7	2340.8	−3455.4	−21,719.7	−5144.6	−3728.5	−6465.1	−122.4	−2377.9
	2012	−71,065.4	−29,397.6	−5893.7	−5109.6	−17,437.0	−28,511.0	−57,020.9	−80,715.3	−4815.5
Somalia	1997	2.1	n.a.	n.a.	n.a.	n.a.	−14.1	184.4	281.2	51.2
	2012	157.3	n.a.	983.6	n.a.	n.a.	n.a.	n.a.	645.4	7.9
Sudan	1997	436.8	n.a.	33.6	n.a.	n.a.	146.1	52.8	−157.6	9.9
	2012	−6.2	364.3	−301.6	−123,064.0	n.a.	3605.5	5584.3	1579.1	1760.2
Uganda	1997	−4953.9	n.a.	−3581.7	4007.1	n.a.	63.3	304.3	171.1	−0.3
	2012	29,598.0	12,217.2	−7013.4	2399.0	−7344.7	4553.5	10,911.5	7318.9	11,227.4
Total	1997	−6124.1	n.a.	n.a.	n.a.	n.a.	n.a.	−9083.7	n.a.	−3739.7
	2012	−99,875.4	n.a.	3782.4	n.a.	n.a.	n.a.	n.a.	−67,975.2	10,635.5

Source: Author's calculations on the basis of UNCTAD (2014)

Table 5 The imports creation effect in terms of product groups (1997 and 2012) (US$ in thousands)

		Food and live animals	Beverages and tobacco	Crude materials. Inedible, except fuels	Mineral fuels, lubricants and related materials	Animal and vegetable oil, fats and waxes	Chemicals and related products, n. e.s.	Manufactured goods	Machinery and transport equipment	Miscellaneous manufactured articles
Djibouti	1997	−1693.3	−0.3	660.5	449.2	n.a.	39.4	310.8	7.0	94.5
	2012	−695.1	54.8	313.3	21.6	5.3	10.6	191.2	716.2	234.2
Eritrea	1997	n.a.	n.a.	n.a.	n.a.	n.a.	n.a.	n.a.	n.a.	n.a.
	2012	−13,601.1	−249.1	−1439.1	−8901.7	−411.3	−969.0	−4528.7	74,364.2	5888.0
Ethiopia	1997	−451.9	−137.7	−313.8	−453.4	−370.8	−2551.2	2452.8	642.0	−718.4
	2012	4294.4	2014.2	5712.0	−81,944.2	−858.6	4610.4	2646.0	17,157.4	5598.2
Kenya	1997	13586.2	100.4	−3896.1	5727.0	n.a.	74.6	157.1	207.8	−102.6
	2012	29455.1	−2274.0	4037.5	761.9	1354.1	685.7	2368.4	15923.4	2493.9
Somalia	1997	−13305.7	8675.4	1428.4	6947.0	−3558.4	−3219.5	2933.7	−386.0	−1554.8
	2012	3080.5	10491.0	21001.9	n.a.	n.a.	1485.9	3313.4	541.3	6508.3
Sudan	1997	−3376.4	338.6	−1128.1	−652.1	441.8	2159.3	160.8	76.2	15.8
	2012	−47034.8	−14591.1	111.6	−48,582.9	−4953.9	5907.5	2629.7	7185.8	1666.9
Uganda	1997	6362.4	−165.0	−3090.3	−37,989.1	−2802.2	6297.3	6876.8	1737.1	6031.2
	2012	7941.8	−8341.8	2978.3	−39,879.9	−1724.5	−21,425.4	−97,963.4	−5597.3	−4999.4
Total	2012	−16559.2	−12896.0	32,715.5	n.a.	n.a.	−9694.3	−91,343.4	110,291.0	17,390.1

Source: Author's calculations on the basis of UNCTAD (2014)

import creation effect in 2012 (almost 29 % decrease in value because of IGAD's new status), but this falloff was driven only by Uganda (Table 5).

3.2 Trade Concentration

The geographical concentration of IGAD members' (except for Sudan and Eritrea) exports and imports was investigated with the Herfindahl-Hirschman Index (HHI). Eritrea experienced the most significant increase in the geographical concentration of exports. In 2000, almost 45 % of Eritrea's exports went to Yemen. By 2012, however, the structure of its main export partners' had shifted towards Canada, where almost 94 % of its exports went. The opposite trend is observed in Djibouti, where exports focused on Ethiopia (47 %) in 1995. In 2012, Djibouti exported mainly to UAE (21 %) and Yemen (18 %). Kenya was characterized by the lowest geographical concentration. Its exports focus was on markets in Uganda, the Netherlands, the UK and USA. Analysing the diversification in import markets we also cannot see any uniform trends in all IGAD states. Only Eritrea and Uganda decreased their import concentrations slightly. In 2012, Eritrea imported primarily from China, Egypt and Italy, while Uganda imported from India, Kenya and China. Kenya and Sudan held the lowest import concentration indexes in 2012. However, import market concentration indices measured in HHI are much lower than HH indexes in exports. This implies a rather moderate concentration of imports of IGAD members (Table 6).

While analysing the synthetic measures of concentration of commodity structures of exports and imports (applying Standard International Trade Classification, Rev.3), we can observe that there is no clear tendency among IGAD member states. States such as Sudan, Somalia and Eritrea were characterized by really high concentration of exports in 2012. In 2012, almost 81 % of Sudan's exports were realized by groups of crude and refined petroleum and gold. Somalia primarily exported livestock (sheep, goats and bovine) and wood charcoal. Eritrea concentrated its exports on gold, silver and hides. Both the Gini coefficient and the Herfindahl Index (HI) in these three countries were high and pointed to the high degree of commodity export concentration. The other IGAD members, especially Kenya, were characterized by more diversified exports in 2012. Kenya exported mainly tea and coffee, cut flowers and refined petroleum. Djibouti concentrated its exports on refined petroleum, livestock and coffee. Ethiopia, in turn, focused on exporting coffee, oily seeds, vegetables and cut flowers. Uganda sold abroad primarily coffee, broadcasting equipment and fish. Uganda is a country that has managed to diversify the structure of its exports most significantly in 1995–2012. In contrast, in Djibouti, Eritrea and Sudan the opposite trend is observed. In the period under analysis Djibouti focused more on exports of livestock and petroleum, while Eritrea exported gold and Sudan sold petroleum and gold abroad. The commodity concentration was lower in imports, regardless of the indicators which were taken into account. Unquestionably, Somalia had the most concentrated imports. In

Table 6 Index of export and import market concentration of the IGAD countries (1995 and 2012)

	Exports		Imports	
	1995	2012	1995	2012
Djibuti	0.273	0.108	0.082	0.118
Eritrea	0.254	0.884	0.112	0.098
Ethiopia	0.134	0.060	0.068	0.099
Kenya	0.064	0.053	0.057	0.063
Somalia	0.541	0.475	0.106	0.160
Sudan	0.068	0.302	0.048	0.062
Uganda	0.076	0.065	0.119	0.080
IGAD	0.045	0.045	0.044	0.060

Note: A HH index below 0.01 indicates a highly competitive index. A HH index below 0.15 indicates an unconcentrated index. A HHI index between 0.15 and 0.25 indicates moderate concentration. A HHI above 0.25 indicates high concentration
HH index formula: $HHI = \Sigma^n S_i^2$, where S_i: market share of country i
Source: Author's calculations on the basis of UNCTAD (2014)

Somalia, vegetables, rice, raw sugar, pasta and wheat flour constituted more than half of the imports value in 2012. The other countries were characterized by more fragmented and diversified imports (Table 7).

3.3 *Trade by Technology Intensity*

A weakness of the export structure of IGAD members is that the high-tech products' share of their exports continues to be low. This opinion is expressed here although it is still uncertain whether in the case of developing countries high-tech exports can really be treated as a solid indication of their technological development (Mani 2000; Srholec 2005). These doubts are usually justified by the statement that technologically advanced exports do not necessarily have to be the result of actual innovative activities in the countries, but rather of a suitable position in global value chains based on revealed comparative advantages (vertical specialization) (Dicken et al. 2011). We can explain the low technological advancements in IGAD countries by their very low share of expenses on research and development. For example, Kenya allocated most expenses to R&D activities, (0.42 % of its GDP in 2012), followed by Uganda, which allocated 0.41 % of its GDP to this goal. In comparison, the average for the EU-27 in this regard was 2.03 % of the GDP in 2012 (Eurostat 2014). Total R&D personnel per million inhabitants amounted to 150 in Ethiopia in 2010, 180 in Kenya, 63 in Uganda in 2007 and 751 in Sudan in 2005[6] (the latest available data). As compared to developed countries, these figures are considerable lower. For instance, in 2010, in UK the total number of R&D personnel per million inhabitants was 8448, in Finland it was 14,900 and in Japan it was 9105. The number of researchers per million dwellers in the IGAD states was also lower. In 2007, there

[6] Overestimated or based on overestimated data according to UNESCO.

Table 7 Synthetic indicators of commodity export and import concentration of the IGAD states (1995 and 2012)

	Exports						Imports					
	Gini coefficient		H index		Share of three most important product groups (%)		Gini coefficient		H index		Share of three most important product groups (%)	
	1995	2012	1995	2012	1995	2012	1995	2012	1995	2012	1995	2012
Djibuti	0.46557	0.625648	0.050	0.099	26.89	46.03	0.431588	0.478608	0.050	0.043	25.91	23.87
Eritrea	0.565295	0.949671	0.152	0.263	57.17	72.85	0.669548	0.610991	0.082	0.050	41.26	30.46
Ethiopia	0.890193	0.823821	0.420	0.168	82.22	64.85	0.59947	0.584886	0.074	0.061	39.00	33.15
Kenya	0.662224	0.528282	0.146	0.088	50.00	43.22	0.571454	0.569173	0.053	0.082	31.58	37.27
Somalia	0.962195	0.885407	0.530	0.476	88.62	80.71	0.629355	0.761722	0.117	0.124	50.40	51.63
Sudan	0.817377	0.786419	0.131	0.335	52.98	83.77	0.556506	0.499137	0.065	0.038	37.10	19.24
Uganda	0.923048	0.560155	0.580	0.087	85.31	38.14	0.558505	0.541522	0.041	0.070	23.17	35.10

Note: H index formula: $HI = \sqrt{\sum_k \left(x_{ik} / \sum_k x_{ik} \right)^2}$, where x_{ik} = country i's exports of product k

Source: Author's calculations using SITC Rev.3 on the basis of UNCTAD (2014)

Table 8 Structure of exports of manufactured products in terms of technological advancements (1995 and 2012)

		Djibouti	Eritrea	Ethiopia	Kenya	Somalia	Sudan	Uganda
Labor-intensive and resource-intensive	1995	23.8 %	18.9 %	97.0 %	35.9 %	16.9 %	83.0 %	17.6 %
	2012	14.0 %	57.8 %	56.1 %	36.0 %	47.7 %	35.6 %[a]	28.5 %
Low-skill and technology-intensive	1995	16.6 %	11.8 %	0.0 %	21.4 %	29.0 %	0.6 %	21.1 %
	2012	11.6 %	1.7 %	2.3 %	14.9 %	1.3 %	11.3 %[a]	15.6 %
Medium-skill and technology-intensive	1995	41.9 %	54.7 %	0.0 %	13.4 %	30.2 %	11.1 %	30.5 %
	2012	41.9 %	9.2 %	31.5 %	16.7 %	31.5 %	11.1 %[a]	17.0 %
High-skill and technology-intensive	1995	17.6 %	14.6 %	3.0 %	29.4 %	23.9 %	5.3 %	30.7 %
	2012	32.5 %	31.3 %	10.0 %	32.4 %	19.5 %	42.0 %[a]	38.9 %

[a]Estimation
Source: UNCTAD

were 30, 93 and 29 researchers in Ethiopia, Kenya and Uganda respectively. As compared to numbers from developed countries (10,094 in Finland, 6942 in Japan, or 6187 in the UK in 2007) these figures are small (UNESCO 2014). In the Global Innovation Index (2013), Uganda was the leader in the IGAD region—it had the 89th place among 142 countries in this ranking, while Sudan held the last but one position in this ranking (World Economic Forum 2013). In turn, in the Knowledge Economy Index 2012, Kenya (111) reached the highest position.

With regard to the share of high-tech product exports in the total manufactured goods' export value, in 2012 it was the highest for Sudan (42 %). However, this data is merely an estimation and this share is questionable because of a lack of latest R&D statistics and the low rank that the country has in the Global Innovation Index. The second country in terms of high-tech exports is Uganda. The country's high percentage of high-tech exports is primarily the result of inflows of foreign direct investment connected with parts and components for electrical and electronic goods and the development of this sector within the country.[7] Also, Uganda has improved its performance partly due to the recently increased investments in oil manufacturing and services sectors, which are reflected in improvements in its exports structure. In Eritrea, Ethiopia and Somalia around half of the manufactured exports still consist of labor-intensive and resource-intensive products.[8] We can see an optimistic tendency in high-skill and technology-intensive exports between 1995 and 2012. The share of these exports increased in all analysed countries, except for Somalia (Table 8).

[7] The World Investment Report 2013 shows that Uganda is the leading recipient of FDI in the East African region (UNCTAD 2013).

[8] The massive oil and gas deposits found in Kenya, Somalia and Uganda, are going to define their economies as so-called petro-dollar economies. In these three countries have operated Pan-Continental, Tullow, Anadarko, ENI, Statoil, CAMAC, BG Group, Swiss Oil, Total, CNOOC, General Energy Plc, Conoco-Philips, Royal Dutch Shell, Exxon Mobil and BP (New African 2014).

4 Competitiveness of Exports of the Selected IGAD States

IGAD's accession to international markets has resulted in adjustments to global trade systems and in changes in competiveness indicators. Generally, we observe economic synergies between Africa and the world by revealing the advantages resulting from combining sectoral and national potentials on a win-win basis. Consequently, it is worth analysing the changes in the competitive positions of IGAD states. The analysis covers three economies in the IGAD region: Ethiopia, Kenya and Sudan and their relations to selected regions and country groups. The rest of the IGAD countries had to be excluded from this study due to lack of reliable and comparable data. Annual export and import dynamics were calculated in order to evaluate the changes in competitive positions. These indicators were adjusted to currency exchange fluctuations (changes of exchange rates of the local currencies to US$) and volatility of commodity prices in the international market. The indicator formula is:

$\Delta_{adj}X^j_n - \Delta X^j_n / \Delta X_n$

where $\Delta_{adj}X^j_n$ is adjusted annual dynamics export of region/country j in n product group,

$\Delta X^j_n / \Delta X_n$ is real dynamics of export value of region/country j in n product group in relation to real annual dynamics of total export of the analysed country.

If the indicator is above zero ($\Delta_{adj}X^j_n - \Delta X^j_n / \Delta X_n > 0$), it means that the analysed country improved its competitive position in a given region/country. If the indicator is below zero ($\Delta_{adj}X^j_n - \Delta X^j_n / \Delta X_n < 0$), it means that the analysed country deteriorated its competitive position in a given region/country.

Generally, it is difficult to find a common tendency in changes in competitive positions in particular product groups. Only in beverages and tobacco all analysed countries indicated improvements in their international positions. In the case of other product groups the results varied widely. For example, Uganda's competitive position improved in almost all product groups and in relation to the analysed regions during 1995–2012. Kenya's competitive position deteriorated in relation to developed countries, especially in relation to Eastern African developing states. In the case of Ethiopia, a lot of data is not available, so it is difficult to point out a trend in changes in its international position. Sudan's competitive position worsened almost in all the product groups in relation to the analysed regions, except for fuels and chemicals (Table 9).

The commodity structure of exports and changes in competitive positions of IGAD countries are related to specialization in international exports. In order to express the relative comparative advantage of a given commodity group in exports, it is most appropriate to calculate the International Specialization Index. For this, we used the revealed comparative advantage index (Balassa 1965). Generally, the IGAD countries export primarily those products in which they have a comparative advantage or in which their competitiveness has improved recently. When we calculate the revealed comparative advantage index for selected countries we see that they moved in the same direction as the changes in competitive positions

Table 9 The matrix of changes in the competitive positions of exports of Ethiopia, Kenya, and Sudan between 1995 and 2012

	Food and live animals	Beverages and tobacco	Crude materials, inedible, except fuels	Mineral fuels, lubricants and related materials	Chemicals and related products, n. e.s.	Manufactured goods	Machinery and transport equipment
Ethiopia							
World	–	+	+	n.a.	–	–	+
Developing countries	+	+	–	–	+	+	n.a.
Developed countries	–	n.a.	+	n.a.	–	–	n.a.
African developing countries	+	+	–	–	+	+	n.a.
Eastern Africa developing countries	+	n.a.	+	n.a.	–	–	+
Kenya							
World	–	+	+	–	+	+	+
Developing countries	+	+	+	–	+	+	+
Developed countries	–	+	+	–	–	–	+
African developing countries	+	+	+	–	+	+	+
Eastern Africa developing countries	+	+	–	–	–	–	–
Sudan							
World	–	+	–	+	+	–	–
Developing countries	–	+	–	+	–	–	–

Developed countries	–	n.a.	–	+	+	–	–
African developing countries	–	n.a.	–	+	+	–	–
Eastern Africa developing countries	–	+	–	+	+	–	–

Note: '+': Improvement in competitive position during 1995–2012; '–': Deterioration of competitive position during 1995–2012
Source: Author's calculations on the basis of UNCTAD (2014)

A. Ethiopia

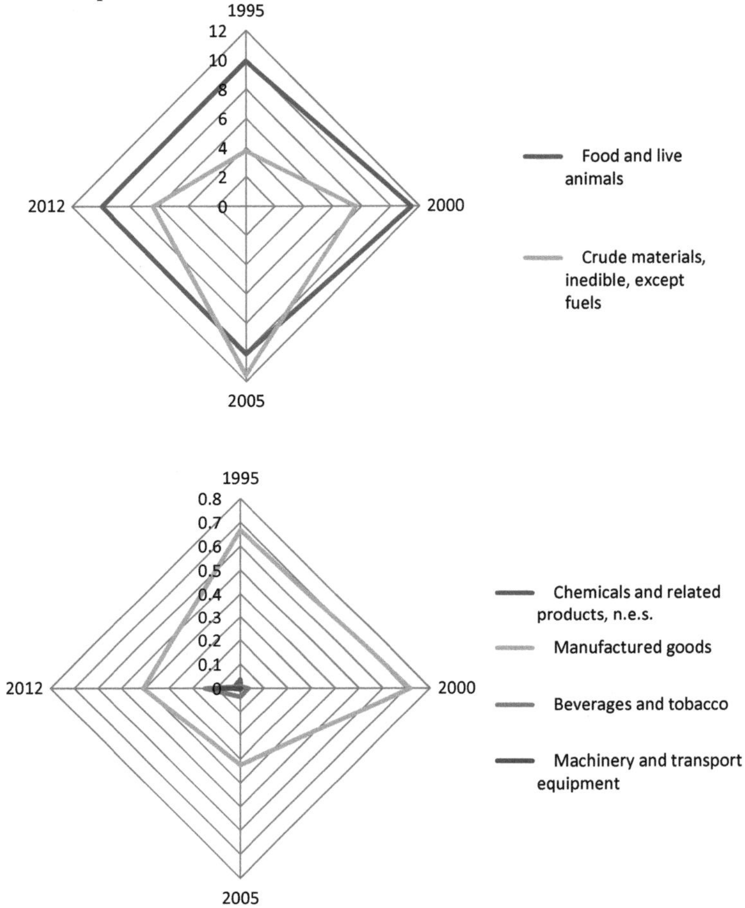

Fig. 8 The revealed comparative advantage indexes (RCA_j) of Ethiopia, Kenya and Sudan between 1995 and 2012. *Source*: Author's calculations on the basis of UNCTAD (2014). *Note*: $RCA_{ik} = (x_{ik}/\Sigma_k\ x_{ik})/(x_{wk}/\Sigma_k\ x_{wk})$, where: x_{ik} = country i's exports of product k; x_{wk} = world exports of product k. $RCA_{jk} > 1$—revealed comparative advantage in export of product j

calculated earlier. Ethiopia and Kenya were characterized by revealed comparative advantages in terms of food and live animals and crude materials. Additionally, Kenya proved to be competitive in beverages and tobacco and in manufactured goods. Sudan specialized only in mineral fuels. The remaining product groups seemed to be still in a disadvantaged position in world trade exchange. None of these countries had comparative advantages in more technologically advanced commodity groups, such as machinery and electronics or transport industry products (Fig. 8).

The following conclusions can be drawn from this analysis for the three selected IGAD members:

B. Kenya

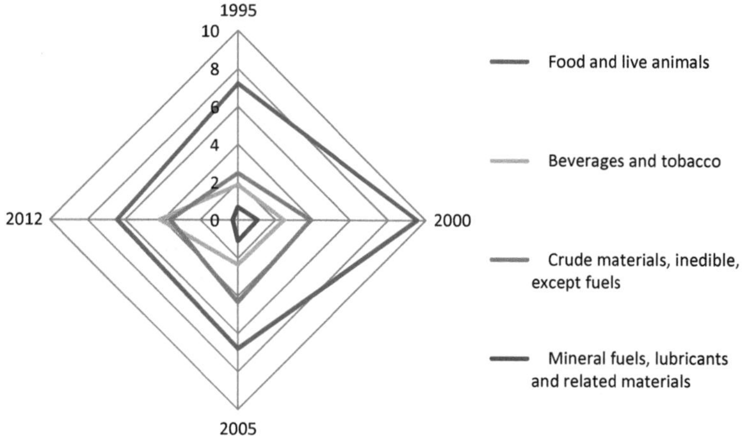

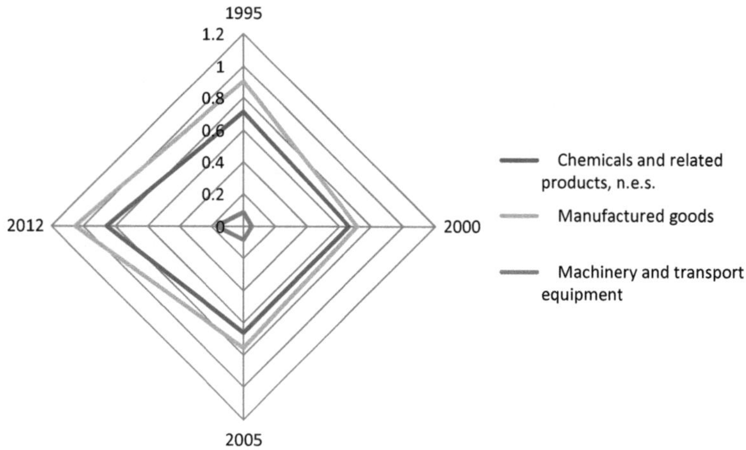

Fig. 8 (continued)

1. Ethiopia:

   (a) generally, we cannot observe any explicit positive trend in Ethiopia's overall competitiveness due to both external and domestic factors including poor international prices, structural supplies and logistical challenges;
   (b) the country improved its competitiveness especially in exports of beverages and tobacco with regard to all analysed regions and groups of states;
   (c) in terms of more advanced goods (for example, machinery and transport equipment), it improved its competitiveness with respect to the world as a whole and its closest neighborhood (East Africa);
   (d) generally, we observe a deterioration in the country's competitive position in terms of minerals fuels, but we do not have complete data to illustrate exports;

C. Sudan

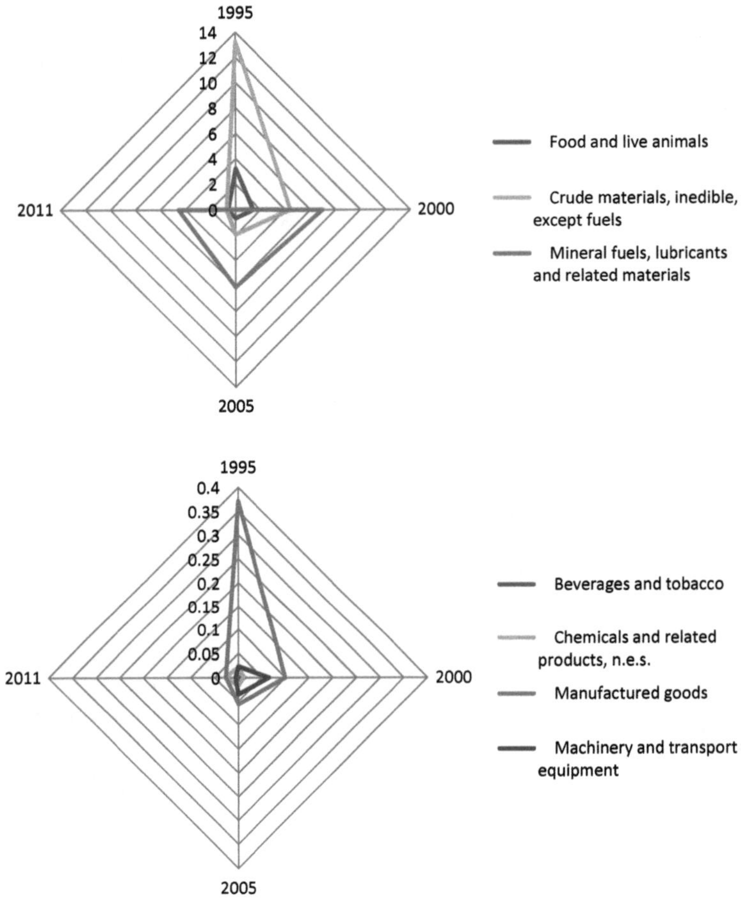

Fig. 8 (continued)

(e) revealed comparative advantages improved in three of the six analysed groups of products between 1995 and 2012. Beverages and tobacco, crude minerals (except fuels) and machinery and transport equipment; food and live animals; chemicals and related products; and manufactured goods deteriorated their RCA_i between 1995 and 2012; the most dramatic decrease in the RCA_i index was noticed in manufactured goods;

(f) in 2012, the country was characterized by revealed comparative advantages in primary products, for example, food, live animals and crude materials; the most significant increase in the RCA_i index was observed in crude materials;

(g) unfortunately more advanced products such as manufactured goods, are still at a comparative disadvantage, though improvements in the RCA_i of these products, especially in machinery and transport equipment have been observed.

2. Kenya:

    (a) we can observe an explicit trend in its competitiveness position: almost all product groups, except for mineral fuels, improved their positions;
    (b) an important aspect of the improvement in Kenya's competitiveness position is the fact that the product group 'machinery and transport equipment' improved its competitiveness in relation to almost all analysed regions (except for Eastern Africa developing countries);
    (c) deterioration in mineral fuels, lubricants and related materials' competitiveness positions with respect to all regions indicates that Kenya's exports have become more advanced;
    (d) between 1995 and 2012 the most significant increase in RCA_i was observed in beverages and tobacco and also in crude minerals, except fuels;
    (e) improvements in Kenya's competitiveness position in exports can be explained by its promoting trade facilitation through enhancements at the Port of Mombasa (now operating 24/7), investing in improvements in the northern corridor road network, starting the construction of the Standard Gauge Railway to link Mombasa to Kampala with a loan from China, launching the Lamu Port and South Sudan-Ethiopia Transport road and rail network with the construction of berths at Lamu and the Jomo Kenyatta International Airport which remains the gateway to East Africa;
    (f) Kenya actively promotes and conducts trade with all its neighbors.

3. Sudan

    (a) we can observe a clear trend in Sudan's competitiveness position: almost all product groups, except for mineral fuels and chemical products, deteriorated their positions; mineral fuels, lubricants and related materials were the only products that improved their competitive positions with regard to all analysed regions; between 1995 and 2012 the most significant increase in RCA_i was observed in mineral fuels, lubricants and related materials;
    (b) deterioration in almost all product groups' competitiveness positions with respect to all regions indicates that Sudan's exports have become less advanced and more concentrated on resource-based goods;
    (c) crude materials, inedible, except fuels significantly deteriorated their RCA_i between 1995 and 2012;
    (d) the pessimistic picture of Sudan's exports' competitiveness and in its comparative advantage is because of the very difficult and unstable political and international situation in this state; although Sudan is a member of COMESA, it has not implemented the tariff-reform program; moreover, the depletion of its foreign exchange reserves limits the potential for a significant reduction of tariffs or other restrictions on imports;

(e) unbalanced exports and concentration on resources is the result of the policy of elimination of external debt.[9]

5 Conclusion

Leaving aside the influence of the redistribution of income between countries with different levels of development, which has already been discussed on numerous occasions in academic literature, the introduction of international fragmentation of production has made it possible to boost selected branches of the IGAD economy in which there had already been comparative advantages or at least a fair chance for increasing international competitiveness (Feenstra 1998). Therefore, it is an attractive development strategy for the Eastern African region to participate in international trade (Ernst & Young 2014).

Reducing the development gap between them and the developed countries and creating a stable economic environment is a priority for the IGAD states (IFC and The World Bank 2014). One of the ways of achieving this goal is by ensuring proper management of foreign trade and gradually joining the global economy. However, IGAD members are very diverse in terms of their economic levels and social and institutional development, and there are also considerable differences in the orientation of their foreign policies which translate into trade flows. The community has not defined one common policy and its members largely realize their particular goals, thus causing even greater economic polarization in the region. However, the omnipresent delocalization and fragmentation of production have not left this region unaffected. The most economically advanced countries in the region joined the international trade the earliest. Transformations in Eastern African economies resulted in foreign enterprises deciding to take advantage of the countries' comparative advantages. Thanks to its factors of production and level of development the IGAD region seems to be attractive for four types of investors, looking for four things: resources, a ready market, a reduction in production costs and strategic assets (for example, the power industry and the railroad network) (Proksch 2003; Ernst & Young 2014).

Nevertheless, we should remember that the IGAD region is still unstable and that it is a pivotal geopolitical pressure point in world politics. That is why economic development and integration within the community may be the solution to political, ideological, religious and ethnic problems which are often compounded by natural disasters like widespread droughts and famines in the region.

[9] As Sudan had become the world's largest debtor to the World Bank and the International Monetary Fund by 1993, its relationship with the international financial institutions soured in the mid-1990s. Since then the government has taken steps to improve its performance. Fiscal expansion, high oil exports and substantial inflows of capital have contributed to a strengthening of the dinar (PRS 2011).

References

Afrika G, Ajumbo G (2012) Informal cross-border trade in Africa: implications and policy recommendations. Africa Economic Brief African Development Bank 3(10)

Agreement Establishing the IGAD (1996) http://www.ifrc.org/docs/idrl/N527EN.pdf. Accessed 15 Jan 2015

Balassa B (1965) Trade liberalisation and “revealed” comparative advantage. Manchester Sch Econ 33:99–123

Balassa B (1998) Trade creation and trade diversion in the European common market. In: Jovanović MN (ed) International economic integration: theory and measurement. Routledge, London, pp 429–525

Cargill T (2010) Our common strategic interests Africa’s role in the post-G8 world. A Chatham House report: June

Chauvin S, Geis A (2011) Who has been affected, how and why? Frankfurt, European Central Bank

Cieślik E (2014) Foreign trade and global value chains of the Southern African Customs Union’s members. Econ Manage Curr Issues Perspect/Ekonomika ir vadyba: aktualijos ir perspektyvos 2(34):93–110

Dicken P, Kelly P, Olds K (2011) Chains and networks, territories and scales: towards a relational framework for analysing the global economy. Glob Netw 1:89–112

Dundas CW (2011) The lag of 21st century democratic elections: in the African Union Member States. Author House

Ernst & Young (2013) Ernst & Young’s attractiveness survey Africa 2013. Ernst & Young

Ernst & Young (2014) Africa 2014 executing growth. Ernst & Young

Eurostat (2014) http://eppeurostateceuropaeu/portal/page/portal/eurostat/home/. Accessed 16 Jan 2015

Feenstra R (1998) Integration of trade and disintegration of production in the global economy. J Econ Perspect 12(4):31–50

Hartzenberg T (2011) Regional integration in Africa WTO. Staff Working Paper (ERSD-2011–14)

Healy S (2013) IGAD and regional security in the Horn. In: Hentz JJ (ed) Routledge handbook of African security. Routledge, London, pp 217–228

Iddrisu MS (2012) The ‘Spaghetti-Bowl’ of Africa’s economic integration: a critique of the African Union’s rationalisation process. LAP Lambert Academic Publishing, Saarbrücken, Germany

IFC, The World Bank (2014) Doing business report 2014. Washington, DC

IGAD (2014) IGAD region. http://www.igadregion.org. Accessed 12 Jan 2015

Mani S (2000) Exports of high-technology products from developing countries: is real or a statistical artefact? UNU/INTECH Discussion Papers

Mataen D (2012) Africa: the ultimate frontier market. Harriman House Ltd, Petersfield

New African (2014) East Africa becomes a petro-dollar region. New African: 72–74

Proksch M (2003) Selected issues on promotion and attraction of foreign direct investment in the least developed countries and economies in transition. Investment Promotion and Enterprise Development Bulletin for Asia and the Pacific 2. UN:1–18

PRS (2011) Political Risk Yearbook: Sudan. Country Report 2011

RIKS Platform (2014) RIKS Database. Accessed 6 Jan 2015

Srholec M (2005) High-tech exports from developing countries: a symptom of technology spurts or statistical illusion. TIK Working Papers (University of Oslo)

The World Bank (2014) The World Bank Open Data. http://data.worldbank.org/. Accessed 12 Jun 2014

UNCTAD (2013) World Investment Report 2013. UNCTAD, Geneva

UNCTAD (2014) UNCTAD statistics. http://unctad.org/en/Pages/Statistics.aspx. Accessed 27 May 2014

UNESCO (2014) UNESCO Institute for Statistics. http://www.uis.unesco.org/Pages/default.aspx. Accessed 28 May 2014
United Nations (2014) UN Comtrade. http://comtrade.un.org/. Accessed 26 May 2014
Weldesellassie I (2011) IGAD as an international organization, its institutional development and shortcomings. J Afr Law 55(1):1–29
World Economic Forum (2013) The Global Competitiveness Report 2013–2014

Currency Union in the East African Community: A Fractional Integration Approach

Hector Carcel, Luis A. Gil-Alana, and Godfrey Madigu

Abstract This study examines inflation rates in the five countries that belong to the East African Community (EAC) and which recently signed a protocol outlining their plans for launching a monetary union within 10 years. Its aim is to examine the persistence in inflation levels. As literature argues about monetary unions, countries hoping to form a union should present similar inflation patterns. Our study shows that the countries present non-mean reversion, confirming that shocks will not recover in the long run. Moreover, fractional co-integration relationships are found between all countries with the exception of Tanzania.

Keywords East Africa union • Monetary union • Fractional integration • Inflation • Persistences

1 Introduction

This study examines the time series behavior of inflation rate levels in five countries (Burundi, Kenya, Rwanda, Tanzania and Uganda) that belong to the East African Community (EAC) which recently signed a protocol outlining their plans for launching a monetary union within 10 years. The study was conducted using a long memory modeling framework based on fractional integration with the aim of analyzing the level of persistence in inflation levels in these Eastern African countries. Alternative measures of persistence based on auto-regressive models were also employed. As most of the literature regarding monetary unions argues, countries hoping to join in such a union should present, among other features, similar inflation patterns the aim of this paper is to find out if this is indeed the case in EAC.

H. Carcel (✉) • L.A. Gil-Alana • G. Madigu
ICS, Navarra Center for International Development, University of Navarra, Pamplona, Spain

Faculty of Economics, University of Navarra, Pamplona, Spain
e-mail: hcarcel@alumni.unav.es; alana@unav.es; gmadigu@alumni.unav.es

A. Heshmati (ed.), *Economic Integration, Currency Union, and Sustainable and Inclusive Growth in East Africa*, Advances in African Economic, Social and Political Development, DOI 10.1007/978-3-319-30432-8_3

Our study shows that the five countries examined present non-mean reverting behavior in their inflation rates, meaning that shocks affecting these countries could be very negative, as they will not return to their original long term projections unless policy measures are implemented. It can, therefore, be suggested that these countries could embark on a monetary union with the aim of diminishing this risk. Additionally, evidence of fractional co-integration is found in bilateral relationships between all the countries that form the EAC with the exception of Tanzania, thus giving further support to a monetary union in the area.

The rest of the paper is organized as follows: Sect. 2 briefly describes the history of the East African countries. Sect. 3 deals with a literature review, while the Sect. 4 presents the data and methodology used for the study. Sect. 5 is devoted to empirical results and Sect. 6 provides a conclusion.

2 A Bit of History

EAC is an intergovernmental organization comprising of five countries in the African Great Lakes region in Eastern Africa: Burundi, Kenya, Rwanda, Tanzania and Uganda. The organization was originally founded in 1967, but it collapsed in 1977 and was officially revived on 7 July 2000. In 2008, after negotiations with the Southern Africa Development Community (SADC) and the Common Market for Eastern and Southern Africa (COMESA), EAC agreed to an expanded free trade area including member states of all three, thus becoming an integral part of the African community. EAC is a potential precursor to the establishment of an East African Federation. In 2010 EAC launched its own common market for goods, labor and capital within the region with the goal of creating a common currency union and eventually a full political federation. In November 2013 a protocol was signed outlining the plans of the five member countries to launch a monetary union within 10 years.

Kenya, Tanzania and Uganda have had a history of cooperation dating back to the early twentieth century. The customs union between Kenya and Uganda in 1917, which the then Tanganyika joined in 1927, was followed by the East African High Commission from 1948 to 1961 and later by the East African Common Services Organization from 1961 to 1967. This was followed by the East African Community until 1977. Inter-territorial cooperation between Kenya colony, the Uganda protectorate and the Tanganyika territory was first formalized in 1948 by the East African High Commission. This provided a customs union and common external tariff, currency and postage; it also dealt with common services in transport and communications, research and education. Following independence, these integrated activities were reconstituted and the High Commission was replaced by the East African Common Services Organization, which many observers thought would lead to a political federation between the three territories. The new organization ran into difficulties because of lack of joint planning and a joint fiscal policy, separate political policies and Kenya's dominant economic position. In 1967 the

East African Common Services Organization was superseded by the East African Community. This body aimed to strengthen ties between members through a common market, a common customs tariff and a range of public services so as to achieve balanced economic growth within the region.

Kenya, Tanzania and Uganda signed a treaty for establishing EAC in 1999, which entered into force in July 2000. In 2007 the treaty was also signed by Burundi and Rwanda, expanding the EAC to five countries. According to the treaty, EAC should first form a customs union, then a common market and a monetary union, and finally a political union. The customs union became operational in 2005 and was formally completed in 2010. The common market protocol was signed in 2009 and the plan is that the creation of a common market, which includes free movement of goods, labor, persons, services and capital, and the right of residence and establishment, will be completed by 2015.

The work of forming a monetary union started early, but proceeded slowly. Thus, in 2007 EAC member countries decided to fast track its establishment and aimed at having it in place by 2012. The intention was to sign a protocol to establish the East African Monetary Union (EAMU) in 2012, which was finally signed in 2013, while actual implementation, though planned to be completed by 2015 is now expected to take several more years. As is evident from the experience of the European Monetary Union (EMU), forming a monetary union is a complicated project and there is non-negligible risk of failure. It is therefore necessary to ensure that the preconditions for forming EAMU are adequate. This entails making sure that economic, political and institutional requirements are in place, since the benefits are likely to be less visible than short-run costs.

3 Literature Review

The Optimum Currency Area (OCA) theory is used to analyze the suitability of a monetary union for a given region. It explores the criteria as well as the costs and benefits of forming a common currency area. The concept of currency areas was founded by Mundell (1961) in his seminal paper 'A Theory of Optimum Currency Areas', followed by Mckinnon (1963) and Kenen (1969). These authors are the founders of the traditional OCA theory, which describes the characteristics that potential monetary union members should possess before they form a single common currency and surrender their national monetary policies and exchange rate adjustments of their national currencies.

When it comes to African monetary unions, most of the literature is related to the current aim of building up a new currency known as the ECO. This currency union of Anglophone West African countries could eventually come to exist in the near future under the name of the West African Monetary Zone (WAMZ). By using fractional integration it has been established that some significant differences exist between these countries. It has been shown, for instance, that shocks to inflation in Sierra Leone are not mean-reverting, while the results for Gambia, Ghana and

Guinea suggest some inflation persistence (Alagidede et al. 2010). Balogun (2007) has proved that independent monetary and exchange rate policies have been relatively ineffective in influencing domestic activities (especially, GDP and inflation) and, therefore, a currency union could benefit the region. After performing a fractional integration analysis on the West African Economic and Monetary Union (WAEMU), Gil-Alana and Carcel (2014) argue that the eight countries that share the CFA as a common currency are tied together not because of their shared economic homogeneity but rather due to their strong historical and traditional ties to France.

Several papers have been published with the purpose of promoting EAC. Using the OCA approach, Mafusire and Brixiova (2013) empirically tested the extent of shock synchronization among EAC members and came to the conclusion that if the countries in the union have major structural differences, a common monetary policy will have differential impacts that may not be helpful to some members. Durevall (2011) has pointed out that EAC has a number of convergence criteria, but these need to be improved upon and revised so that the union succeeds, and Kishor and Ssozi (2009) have found that the proportion of shocks which are common across different countries is small, implying weak synchronization, with this degree of synchronization having improved after the signing of the EAC treaty in 1999. Several authors have studied the viability of a monetary union in EAC. These studies have used different models and have reached different conclusions. For example, Buigut and Valev (2005) applied a two-variable SVAR model to test for shock correlation in EAC countries; they found that forming a monetary union in EAC is not feasible. Conversely, Mkenda (2001) and Falagiarda (2010) employed the G-PPP approach which uses a co-integration analysis to conclude that a monetary union in East Africa could be a viable option. Lastly, Sheikh et al. (2011) and Opolot and Osoro (2004) studied the feasibility of forming a monetary union in EAC using the business cycle synchronization approach of Hodrick-Prescott and Baxter-King filters. They found a low degree of synchronization between EAC members, but with improved results in recent years.

Some studies were conducted on macroeconomic issues in African countries using long memory techniques. For instance, focusing on exchange rates, Mokoena et al. (2009) show that except for South Africa, none of the SADC real exchange rates are fractionally integrated. Fractional integration has also been used to analyze the stock market structure (Anouro and Gil-Alana 2010; Rambaccussing 2010), inflation (Gil-Alana and Barros 2013) and housing prices (Gil-Alana et al. 2013) in African countries. However, none of these studies have focused on EAC. With this work we attempt to contribute to literature on the formation of a monetary union in EAC by conducting a long memory fractional integration analysis of the persistence of the inflation rates in five East African countries.

4 Data and Methodology

We used monthly data of the inflation rate from January 2004 to December 2013 in the five EAC member countries. The data was obtained from the 'Trading Economics' website. With these time series we conducted a long memory and fractional integration analysis. As we can see in Fig. 1, the five graphs show considerable volatility in inflation rates, each of them presenting a relatively different form, but with constant ups and down throughout the period that is analyzed.

Long memory is a feature of the data which tells us that observations that are far from one another in time are highly correlated. Based on the frequency domain, we can provide the following definition of long memory: Let us suppose that $\{x_t, t=0, \pm 1, \ldots\}$ is a covariance stationary process with auto-covariance function $E(x_t - Ex_t)(x_{t-j} - Ex_t) = \gamma_j$, such that it has an absolutely continuous spectral distribution function, with a spectral density function denoted by $f(\lambda)$, and defined as:

$$f(\lambda) = \frac{1}{2\pi} \sum_{jÇ-\infty}^{j=\infty} \gamma_j \cos \lambda j, \qquad -\pi < \lambda \leq \pi. \tag{1}$$

Then, x_t displays the property of long memory if the spectral density function has a pole at some frequency λ in the interval $[0, \pi)$, that is:

$$f(\lambda) \to \infty, \quad as\ \lambda \to \lambda^*, \quad \lambda^* \in [0, \pi). \tag{2}$$

Empirical literature has mainly focused on the case where the singularity or pole in the spectrum occurs at the zero frequency, that is, $(\lambda^* = 0)$. This is the standard case of I(d) models of the form:

$$(1 - L)^d x_t = u_t, \quad t = 0, \pm 1, \ldots, \tag{3}$$

where d can be any real value, L is the lag-operator $(Lx_t = x_{t-1})$ and u_t is I(0), defined for our purposes as a covariance stationary process with a spectral density function that is positive and finite at the zero frequency. The polynomial $(1 - L)^d$ in Eq. (3) can be expressed in terms of its binomial expansion, such that, for all real *d*:

$$(1 - L)^d = \sum_{j=0}^{\infty} \Psi_j L^j = \sum_{j=0}^{\infty} \binom{d}{j} (-1)^j L^j = 1 - dL + \frac{d(d-1)}{2} L^2 - \ldots, \tag{4}$$

and thus:

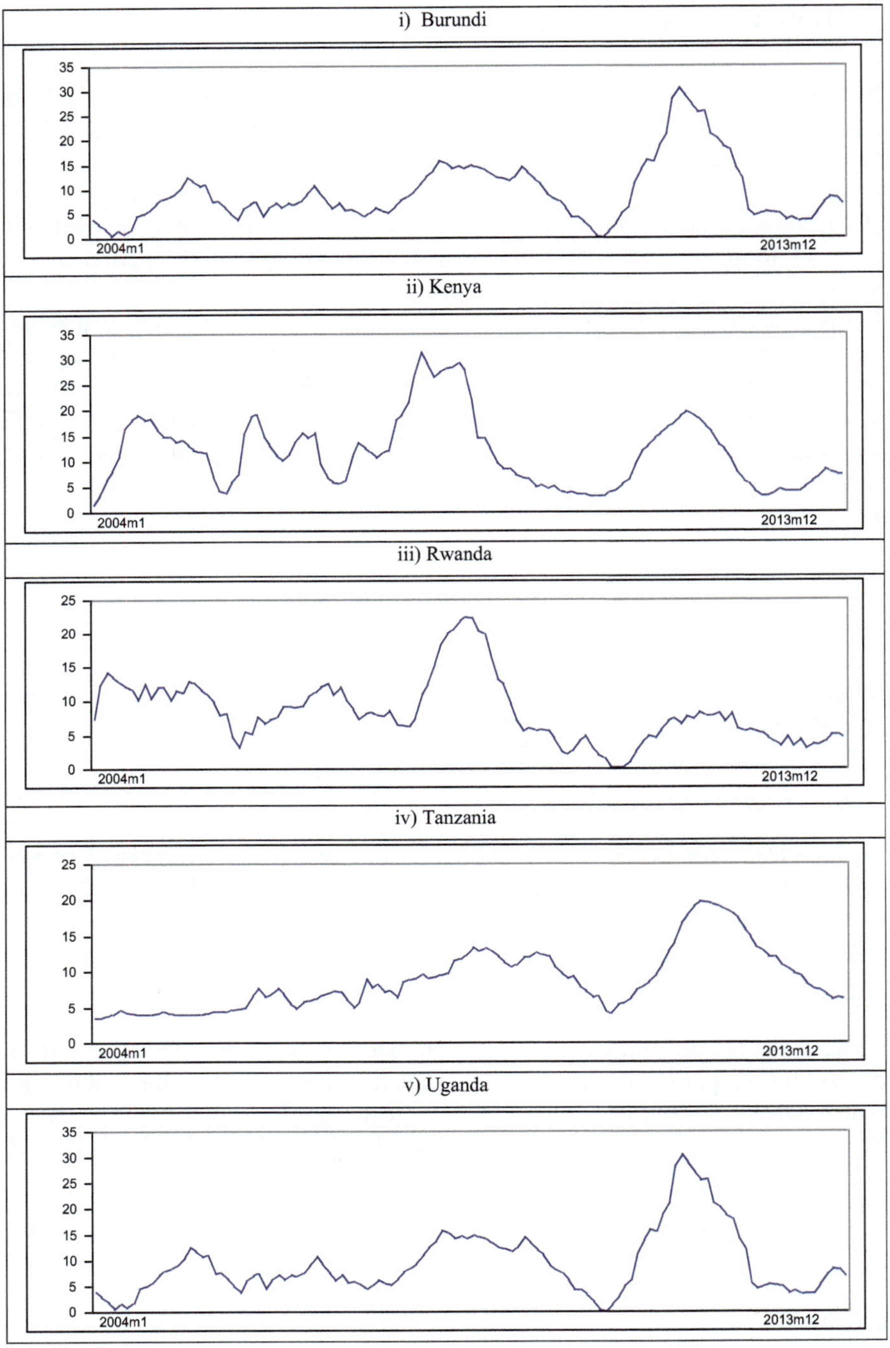

Fig. 1 Original time series

$$(1 - L)^d x_t = x_t - dx_{t-1} + \frac{d(d-1)}{2} x_{t-2} - \ldots,$$

implying that Eq. (3) can be expressed as:

$$x_t = dx_{t-1} - \frac{d(d-1)}{2} x_{t-2} + \ldots + u_t.,$$

In this context, d plays a crucial role since it indicates the degree of dependence of the time series: the higher the value of *d*, the higher the level of association will be between the observations (Barros et al. 2011). This process also admits an infinite moving average (MA) representation such that:

$$x_t = \sum_{k=0}^{\infty} a_k u_{t-k},$$

where:

$$a_k = \frac{\Gamma(k+d)}{\Gamma(k+1)\Gamma(d)},$$

and $\Gamma(x)$ represents the Gamma function.

Given the parameterization in Eq. (3) we can distinguish several cases of interest depending on the value of d. Thus, if $d = 0$, $x_t = u_t$, x_t is said to be 'short memory' or I(0), and if the observations are auto-correlated (that is, AR), then they are 'weak' in form, in the sense that the values in the auto-correlations are decaying exponentially; if $d > 0$, x_t is said to be 'long memory', so named because of the strong association between observations which are distant in time. Here, if d belongs to the interval (0, 0.5) x_t is still covariance stationary, while $d \geq 0.5$ implies non-stationarity. Finally, if $d < 1$, the series is mean reverting in the sense that the effect of the shocks disappears in the long run, contrary to what happens if $d \geq 1$ with shocks persisting forever.

Two univariate methods of estimation of the fractional differencing parameter were employed: one is a Whittle parametric approach in the frequency domain (Dahlhaus 1989), while the other is a semi-parametric 'local' Whittle approach (Robinson 1995). In addition, a simple AR(1) model was also considered as an alternative approach to measure persistence by means of the AR coefficient.

In a multivariate framework, the bivariate relationships between the variables were examined by means of fractional co-integration; two methods were employed here, the one based on Engle and Granger's (1987) methodology and extended to the fractional case by Gil-Alana (2003), and a Hausman test for the null of no co-integration against the alternative of fractional co-integration, as developed by Marinucci and Robinson (2001).

5 Empirical Results

The first thing we did was to check the persistence of the series by conducting a simple AR(1) model for each case. The estimated AR coefficients, displayed in Table 1 are in all cases extremely close to 1, ranging from 0.958 in the case of Kenya to 0.993 for Tanzania. This clearly indicates that the series are highly persistent and possibly non-stationary. Thus, we also conducted various unit root procedures. In particular, we used the ADF (Dickey and Fuller 1979) and Phillips and Perron (PP 1988) approaches. These are some of the most commonly employed tests in literature.

As expected, the p-values reveal that we cannot reject the null hypothesis of a unit root in any of the series (Table 2). Therefore, according to these results, we should take first differences in order to make them I(0) stationary. Nevertheless, we should take into account that these unit root tests might have very low power when directed against specific alternatives such as trend-stationary models (DeJong et al. 1992), structural breaks (Campbell and Perron 1991), regime-switching (Nelson et al. 2001) or fractionally integration (Diebold and Rudebusch 1991; Hassler and Wolters 1994; Lee and Schmidt 1996). In what follows, we focus on the latter type of alternatives, noting that fractional integration includes the classic unit root models as particular cases of interest.

We estimated the fractional differencing parameter d in a model given by the following form:

$$y_t = \alpha + \beta t + x_t; \quad (1 - L)^d x_t = u_t, \quad t = 1, 2, \ldots., \tag{5}$$

where y_t is the observed time series for each country, α and β are the unknown coefficients corresponding to an intercept and a linear trend, and the resulting

Table 1 Estimated AR(1) coefficients

	Burundi	Kenya	Rwanda	Tanzania	Uganda
Estimated AR(1)	0.9747	0.9588	0.9774	0.9933	0.9705

Table 2 Unit root test results

ADF	Burundi	Kenya	Rwanda	Tanzania	Uganda
No regressors	0.3922	0.3337	0.06076	0.391	0.4692
Intercept	0.117	0.2469	0.2986	0.4051	0.5100
Time trend	0.3593	0.4745	0.5007	0.8608	0.3685
PP	Burundi	Kenya	Rwanda	Tanzania	Uganda
No regressors	0.1559	0.1997	0.2049	0.4111	0.1654
Intercept	0.0981	0.0802	0.1847	0.2921	0.1096
Time trend	0.2817	0.1575	0.2090	0.6886	0.3212

Note: The values in the tables refer to the corresponding p-values

Table 3 Estimates of d and 95 % confidence intervals

	No regressors	An intercept	A linear trend
(i) White noise disturbances			
Burundi	**0.98 (0.87, 1.12)**	**1.02 (0.91, 1.18)**	**1.02 (0.91, 1.18)**
Kenya	1.44 (1.28, 1.65)	1.44 (1.28, 1.66)	1.44 (1.27, 1.66)
Rwanda	1.24 (1.12, 1.39)	1.20 (1.09, 1.35)	1.20 (1.09, 1.35)
Tanzania	1.24 (1.13, 1.38)	1.27 (1.16, 1.40)	1.27 (1.16, 1.40)
Uganda	1.33 (1.21, 1.48)	1.37 (1.24, 1.53)	1.37 (1.24, 1.53°)
(ii) Bloomfield disturbances			
Burundi	**0.97 (0.69, 1.32)**	**0.96 (0.68, 1.32)**	**0.96 (0.67, 1.33)**
Kenya	**1.04 (0.79, 1.42)**	**1.03 (0.68, 1.41)**	**1.03 (0.72, 1.39)**
Rwanda	**1.27 (0.96, 1.69)**	**1.32 (0.98, 1.74)**	**1.33 (0.98, 1.77)**
Tanzania	1.31 (1.03, 1.62)	1.34 (1.11, 1.66)	1.34 (1.11, 1.66)
Uganda	**1.26 (0.97, 1.58)**	**1.26 (0.97, 1.60)**	**1.26 (0.97, 1.61)**
(iii) Monthly AR(1) disturbances			
Burundi	**0.97 (0.88, 1.10)**	**1.03 (0.93, 1.17)**	**1.03 (0.93, 1.17)**
Kenya	1.42 (1.25, 1.65)	1.42 (1.25, 1.65)	1.40 (1.24, 1.64)
Rwanda	1.25 (1.13, 1.42)	1.19 (1.07, 1.36)	1.19 (1.07, 1.36)
Tanzania	1.19 (1.09, 1.34)	1.24 (1.14, 1.38)	1.24 (1.14, 1.37)
Uganda	1.26 (1.16, 1.40)	1.31 (1.20, 1.46)	1.32 (1.20, 1.46)

Note: In bold, evidence of unit roots (d = 1) at the 5 % level

errors, x_t, are supposed to be I(d). First we employed a parametric approach, and thus, we needed to specify a functional form for the d-differenced process. We considered three different cases corresponding to white noise disturbances, Bloomfield-type and monthly seasonal AR. Bloomfield's (1973) model is a non-parametric approach that produces auto-correlations decaying exponentially as in the AR(MA) case. Finally, we used monthly AR(1) disturbances based on the monthly nature of the series examined.

Table 3 gives the estimated values of d under three different specifications assuming (a): no deterministic terms [that is, $\alpha = \beta =$ in Eq. (5)], (b): an intercept (α unknown and $\beta = 0$), and (c): an intercept with a linear time trend (α and β unknown).

The most noticeable feature observed in Table 3 is that evidence of mean reversion (that is, $d < 1$) is not found in any single case since all the confidence intervals include values of d equal to or higher than 1. For white noise and monthly AR(1) u_t, the unit root null cannot be rejected for the case of Burundi; however, this hypothesis is decisively rejected in favor of $d > 1$ in the remaining four series. If u_t follows Bloomfield's (1973) non-parametric exponential model this explosive behavior is attached in the case of Tanzania but the unit root null cannot be rejected in any of the other four series.

In Table 4 we use a semi-parametric method proposed by Robinson (1995) and modified later by Abadir et al. (2007) among many others. This is a 'local' Whittle estimator in the frequency domain, using a band of frequencies that degenerates to

Table 4 Estimates of d based on the Whittle semi-parametric method

	10	11	12	15
Burundi	**1.095**	1.311	1.288	**0.939**
Kenya	1.358	1.457	1.342	**0.902**
Rwanda	**0.888**	**0.933**	**1.059**	**1.110**
Tanzania	**1.209**	1.335	1.476	1.253
Uganda	1.408	1.500	1.500	1.210
95 % lower int.	0.739	0.752	0.765	0.787
95 % upper int.	1.260	1.247	1.237	1.212

Note: In bold, evidence of unit roots ($d = 1$) at the 5 % level

zero. The results here are consistent with the parametric ones and evidence of unit roots or values of d above 1 are obtained in all cases. This indicates that the series are not mean reverting, implying that if there is a negative shock, the series will not recover by themselves in any single country to their original long term projections, and strong policy measures should be implemented to recover the original values. In addition to this, the fact of having similar results across countries gives further support to the idea that a currency union could be beneficial for these countries.

Next we focused on the bivariate relationships between the countries and the first thing we did was to employ Engle and Granger's (1987) methodology to test for co-integration. We used this methodology rather than Johansen's (1988, 1991) system-based one since later we will extend Engle and Granger's method to the fractional case. The results using this approach are given in Table 5.

Using this approach, we only found evidence of no co-integration in the case of Tanzania with respect to the other four countries. In the rest of the cases, p-values above 0.05 indicated that we reject the null of no co-integration against the alternative of co-integration. However, as mentioned earlier with respect to the unit root procedures, this method can be seriously biased due to the fact that no fractional alternatives have been taken into account.

In what follows we use the methodology developed by Gil-Alana (2003), testing the null of no co-integration against the alternative of fractional co-integration. However, in a bivariate context, a necessary condition for co-integration is that the two individual series must be co-integrated of the same order. Thus, as a preliminary screening, we tested the homogeneity condition of the order of integration in the bivariate systems (that is, Ho: $d_x = d_y$), where d_x and d_y are the orders of integration of the two individual series, by using an adaptation of the Robinson and Yajima (2002) statistic to the Whittle semi-parametric estimation method. The results, though not reported, indicated that all series displayed similar integration orders.

Two different approaches were employed to test for the possibility of fractional co-integration. First, following Gil-Alana (2003) we tested the null of no co-integration against the alternative of fractional co-integration in the estimated residuals from the regression of one country against another. Here, we considered the two cases of uncorrelated (white noise) errors (in Table 6) and correlated

Table 5 Engle and Granger's (1987) co-integration results

	Burundi	Kenya	Rwanda	Tanzania	Uganda
Burundi	–	–	–	–	–
Kenya	0.11860	–	–	–	–
Rwanda	0.23370	0.08135	–	–	–
Tanzania	0.04972[a]	0.04570[a]	0.03636[a]	–	–
Uganda	0.07771	0.15270	0.14660	0.00819[a]	–

[a]Evidence of no co-integration at the 5 % level

Table 6 Gil-Alana's (2003) method of fractional co-integration with uncorrelated errors

	Burundi	Kenya	Rwanda	Tanzania	Uganda
Burundi	–	–	–	–	–
Kenya	(0.89, 1.19)	–	–	–	–
Rwanda	(0.80, 1.13)	(1.11, 1.52)			–
Tanzania	(0.86, 1.15)	(1.25, 1.63)	(1.08, 1.36)		–
Uganda	(0.85, 1.18)	(1.19, 1.63)	(1.04, 1.32)	(0.94, 1.21)	–

(Bloomfield) disturbances (in Table 7). Then, we also considered the Hausman test for Marinucci and Robinson's (2001) no co-integration.

Starting with Gil-Alana (2003) and assuming white noise disturbances (Table 6) we observe that the estimated degree of differentiation is within the unit root interval or even in some cases above 1, thus finding no evidence of co-integration of any degree. However, allowing for auto-correlated disturbances (Table 7) we observe a reduction in the degree of integration of the series and, though the unit root null hypothesis cannot be rejected, we observe many values below 1 suggesting some degree of fractional co-integration.

Next we performed Marinucci and Robinson's (2001) Hausman test. This method is based on:

$$H_{is} = 8\,s\left(\hat{d}^{*} - \hat{d}_{i}\right)^{2} \to_{d} \chi_{1}^{2} \qquad \frac{1}{s} + \frac{s}{T} \to 0, \tag{6}$$

where i = x, y refers to each of the series under examination, s is the bandwidth number (in our case we chose $s = (T)^{0.5}$), $\hat{d}_i$ are the univariate estimates of the parent series and $\hat{d}^{*}$ is a restricted estimate obtained in the bivariate representation under the assumption that $d_x = d_y$. The results using this approach are given in Table 8.

For some of the bivariate relationships we found some evidence of co-integration, in particular for Burundi-Kenya, Burundi-Rwanda, Burundi-Uganda, Kenya-Rwanda and Kenya-Uganda. These results, together with the non-co-integration results obtained for Tanzania with the rest of the countries obtained with the Engle-Granger test show that there has traditionally been more co-integration in inflation levels between Burundi, Kenya, Rwanda and Uganda

Table 7 Gil-Alana's (2003) method of fractional co-integration with correlated errors

	Burundi	Kenya	Rwanda	Tanzania	Uganda
Burundi	–	–	–	–	–
Kenya	(0.37, 1.11)	–	–	–	–
Rwanda	(0.44, 1.08)	(0.44, 1.32)			–
Tanzania	(0.61, 1.30)	(0.72, 1.40)	(0.97, 1.76)		–
Uganda	(0.50, 1.16)	(0.57, 1.24)	(0.89, 1.69)	(0.75, 1.37)	–

Table 8 Hausman test of no co-integration

	Burundi	Kenya	Rwanda	Tanzania
Kenya	$H_{as} = 15.093$[a] $H_{as} = 4.319$[a] **$d^{a} = 0.682$**	–		
Rwanda	$H_{as} = 12.323$[a] $H_{as} = 10.071$[a] **$d^{a} = 0.722$**	$H_{as} = 0.236$ $H_{as} = 3.928$[a] **$d^{a} = 0.855$**	–	
Tanzania	$H_{as} = 2.683$ $H_{as} = 1.209$ $d^{a} = 0.925$	$H_{as} = 1.220$ $H_{as} = 3.130$ $d^{a} = 1.026$	$H_{as} = 3.769$ $H_{as} = 1.079$ $d^{a} = 1.322$	–
Uganda	$H_{as} = 8.152$[a] $H_{as} = 15.458$[a] **$d^{a} = 0.792$**	$H_{as} = 0.089$ $H_{as} = 10.131$[a] **$d^{a} = 0.877$**	$H_{as} = 2.871$ $H_{as} = 0.078$ $d^{a} = 1.246$	$H_{as} = 3.130$ $H_{as} = 3.163$ $d^{a} = 1.021$

[a]We reject the null of no co-integration in favor of fractional co-integration at the 5 % level

than with Tanzania. The lack of co-integration for Tanzania can be explained by the fact that for many years this country has kept stronger economic and trading ties with SADC, an intergovernmental organization to which Burundi, Kenya, Rwanda and Uganda do not belong.

6 Conclusion

We conducted a long memory and fractional integration analysis on inflation rate levels in the five member countries of the East African Community: Burundi, Kenya, Rwanda, Tanzania and Uganda. We did so to examine if these countries share similar characteristics for future economic integration.

We first tested persistence by considering a simple AR(1) process and the AR coefficient was found to be very close to 1 in the five series examined. Due to this high degree of persistence, we also performed unit root tests and the results based on the ADF (1979) and the PP (1988) methods again support the non-stationary I (1) specification in the five countries examined. This result however could be questioned due to the low power of the unit root procedures if the series are fractionally integrated. Using this latter approach, the series displayed orders of integration above 1, this being a clear sign that inflationary shocks in these countries will take permanent effect. Moreover, looking at bivariate fractionally co-integration relationships among the countries, the results support this hypothesis in all countries except Tanzania. This can be explained by the fact that for many years Tanzania has kept stronger economic and trading ties with SADC, to which Burundi, Kenya, Rwanda and Uganda do not belong. In addition, during the last decade, Tanzania has also received immense foreign direct investment in its mining industries, mainly gold, whereas in the other four countries agriculture has remained the main economic sector. We believe that it will be essential to take these features into account as monetary unions usually require homogeneity in inflation levels in country members. In addition, it can also be argued that joining a monetary union will provide more macroeconomic and financial stability to the region, thus leading to possible lower levels of inflationary persistence in the future.

References

Abadir KM, Distasoand W, Giraitis L (2007) Nonstationarity-extended local Whittle estimation. J Econometrics 141:1353–1384

Alagidede P, Colemanand S, Cuestas JC (2010) Persistence of inflationary shocks: implications for West African monetary union. Sheffield Economic Research Paper Series SERP Number: 2010020

Anouro E, Gil-Alana LA (2010) Mean reversion and long memory in African stock market prices. J Econ Finance 35(3):296–308

Balogun ED (2007) Monetary policy and economic performance of West African monetary zone countries. Munich Personal RePEc Archive Paper No. 4308

Barros CP, Gil-Alana LA, Payne JE (2011) An analysis of oil production by OPEC countries: persistence, breaks, and outliers. Energy Policy 39:442–453

Bloomfield P (1973) An exponential model in the spectrum of a scalar time series. Biometrika 60:217–226

Buigut SK, Valev NT (2005) Is the proposed East African monetary union an optimal currency area? A structural vector autoregression analysis. World Dev 33(12):2119–2133

Campbell JY, Perron P (1991) Pitfalls and opportunities: what macroeconomists should know about unit roots. NBER Macroecon Annu 6:1141–1201

Dahlhaus R (1989) Efficient parameter estimation for self-similar process. Ann Stat 17:1749–1766

DeJong DN, Nankervis JC, Savinand NF, Whiteman CH (1992) The power problem of unit root tests in time series with autoregressive errors. J Econometrics 53:323–343

Dickey DA, Fuller WA (1979) Distribution of the estimators for autoregressive time series with a unit root. J Am Stat Assoc 74:427–431

Diebold FS, Rudebusch G (1991) On the power of Dickey-Fuller tests against fractional alternatives. Econ Lett 35:155–160

Durevall D (2011) East African community: pre-conditions for an effective monetary union. Working Papers in Economics, University of Gothenburg, no. 520

Engle RF, Granger CWJ (1987) Cointegration and error correction: representation, estimation, and testing. Econometrica 55:251–276

Falagiarda M (2010) Are the East African countries ready for a common currency? Traditional indicators and cointegration analysis. Riv Politic Economica 10:153–204

Gil-Alana LA (2003) Testing of fractional cointegration in macroeconomic time series. Oxf Bull Econ Stat 65(4):517–529

Gil-Alana LA, Barros CP (2013) Inflation forecasting in Angola. A fractional approach. Afr Dev Rev 85(1):91–104
Gil-Alana LA, Carcel H (2014) Fractional integration in the West African economic and monetary union. Afr J Econ Sustain Dev 3:179–199
Gil-Alana LA, Gupta R, Aye GC (2013) Testing for persistence with breaks and outliers in South African house prices. J R Estate Lit 21(2):293–314
Hassler U, Wolters J (1994) On the power of unit root tests against fractional alternatives. Econ Lett 45:1–5
Johansen S (1988) Statistical analysis of cointegration vectors. J Econ Dynam Contr 12:231–254
Johansen S (1991) Estimation and hypothesis testing of cointegration vectors in Gaussian vector autoregressive models. Econometrica 59:1551–1580
Kenen PB (1969) The optimum currency area: an eclectic view. In: Mundell R, Swoboda A (eds) Monetary problems of the international economy. University of Chicago Press, Chicago, IL, pp 42–60
Kishor NK, Ssozi J (2009) Is the East African community an optimum currency area? MPRA Paper No. 17645
Lee D, Schmidt P (1996) On the power of the KPSS test of stationarity against fractionally integrated alternatives. J Econometrics 73:285–302
Mafusire A, Brixiova Z (2013) Macroeconomic shock synchronization in the East African community. Global Econ J 13(2):261–280
Marinucci D, Robinson PM (2001) Semiparametric fractional cointegration analysis. J Econometrics 105:225–247
McKinnon RI (1963) Optimum currency areas. Am Econ Rev 53(4):717–725
Mkenda BK (2001) Is East Africa an optimum currency area? Göteberg University Department of Economics Working Paper, 41
Mokoena TM, Gupta R, Van Eyden R (2009) Testing for fractional integration in southern African development community real exchange rates. S Afr J Econ 77(4):531–537
Mundell RA (1961) A theory of optimum currency areas. Am Econ Rev 51(4):657–665
Nelson CR, Pigerand J, Zivot E (2001) Markov regime-switching and unit root tests. J Bus Econ Stat 19:404–415
Opolot J, Osoro N (2004) Is the East African community suitable for a monetary union? An enquiry of idiosyncrasies and synchronization of business cycles. Bank of Uganda, Kampala, Uganda
Phillips PCB, Perron P (1988) Testing for a unit root in time series regression. Biometrica 75:335–346
Rambaccussing D (2010) Fractional integration, return predictability and unconditional risk: evidence from African stock markets. Afr Rev Econ Finance 1(2):72–89
Robinson PM (1995) Gaussian semi-parametric estimation of long range dependence. Ann Stat 23:1630–1661
Robinson PM, Yajima Y (2002) Determination of cointegrating rank in fractional systems. J Econometrics 106:217–241
Sheikh KA, Azam MN, Rabby TG, Alamand GM, Khan I (2011) Monetary union for the development process in the East African community: business cycle synchronization approach. Afr J Bus Manag 5(17):7632–7641

Part II
Financial Sector Development and Common Currency

Financial Sector Development–Economic Growth Nexus in Rwanda

Caleb Tamwesigire, Thomas Bwire, and Pascal Munyankindi

Abstract This paper investigates the links between a number of relevant financial indicators and economic growth in Rwanda over the period 2006Q1–2014Q4. A triangulation of the Vector Error Correction Model (VECM) and Granger non-causality approaches and impulse response functions are employed to investigate the dynamic inter-relationship between various measures of financial sector development and economic growth and for addressing questions of the direction of causality between financial sector development and growth in the country. Empirical results suggest that broad money (M3)—a target metric that Rwandan monetary authorities use in influencing the level of economic activities in the economy—has a positive significant long-run correlation with GDP and that causality runs both ways. This confirms the existence of a nexus between financial sector development and economic growth in Rwanda. An impulse response analysis reveals that monetary policy actions in Rwanda are reactionary. The policy rate responds only in the fifth quarter when overheating in the economy following an initial shock in money supply has occurred. This suggests that to anchor expectations of economic agents and for maintaining monetary policy credibility, a forward looking monetary policy framework could be warranted.

Keywords Financial development • Economic growth • Causality tests • Vector auto-regressive • Rwanda

---

C. Tamwesigire
College of Business and Economics, University of Rwanda, Kigali, Rwanda
e-mail: ctamwesigire@yahoo.com; ctamwesigire@sfb.ac.ur

T. Bwire (✉)
Bank of Uganda, Kampala, Uganda
e-mail: tbwire@bou.or.ug; bwiret@yahoo.co.uk

P. Munyankindi
National Bank of Rwanda, Kigali, Rwanda
e-mail: pmunyankindi@bnr.rw

A. Heshmati (ed.), *Economic Integration, Currency Union, and Sustainable and Inclusive Growth in East Africa*, Advances in African Economic, Social and Political Development, DOI 10.1007/978-3-319-30432-8_4

1 Introduction

Financial markets are known to be vital for stimulating real economic activity. A well-developed financial market also efficiently mobilizes savings and channels them into investments, which are growth enhancing (McKinnon 1973; Rousseau and Sylla 2001; Shaw 1973). As argued in Schumpeter (1932), entrepreneurs require credit to produce which implies that economic growth requires investments and for the realization of that investment, credit services are necessary. Moreover, Gurley and Shaw (1955) suggest that financial intermediation stimulates economic growth by improving resource allocation and investment opportunities.

Following this, since the 1990s the government and the National Bank of Rwanda have been implementing reforms in the financial system with the objective of encouraging competition, increasing domestic resource mobilization, enhancing operational efficiency and ensuring the stability of the financial system. These reforms have generated positive results in many important respects, including in particular, a rapid expansion of the banking system (financial deepening) and an even faster expansion of bank lending to the private sector. The ratio of broad money (M3) to GDP rose from 17.1 % in 2004–2005 to 23 % in 2013–2014. Over the same period, the ratio of private sector credit to GDP increased from 3.5 to 14 %. The supply of bank credit to the private sector increased approximately fourfold in real terms (that is, after adjusting for inflation) since the start of the millennium. Notwithstanding this remarkable expansion, the financial system remains relatively underdeveloped, not just in terms of its small size in relation to GDP but also in terms of its lack of diversity.

As with most other low income African countries, Rwanda's financial system is dominated by commercial banks which hold about 66.8 % of the total assets of the financial system; these are followed by pensions (17.5 %), insurance (9.9 %) and micro-finance institutions (5.9 %). As a result, most firms seeking finance for investments in the domestic market have to rely on loan finance for which the most important source is the banking system. Despite the fact that Rwanda's financial systems are still quite shallow, the economy has achieved robust and sustained growth over the last 10 years, averaging about 8 % in real terms.

Whether financial market development underlies Rwanda's robust economic performance remains a subject of speculation. It may well be the case that in addition to financial development, there are other structural factors which are important for stimulating long-run economic growth, in which case it may be argued that financial development is a necessary but not a sufficient condition for stimulating economic growth. Also, as argued by Patrick (1966), in Rwanda not only is financial intermediation supply-leading but it is also demand-following. Thus, the question we address using a triangulation of the Vector Error Correction Model (VECM) and Granger non-causality approaches and impulse response functions is whether financial sector development is growth enhancing in Rwanda. More specifically, this paper examines the long-run relationship between indicators of financial development and economic growth and detects the direction of causality between these variables. Our empirical results for the sample period suggest that broad money (M3) has a positive significant long-run correlation with GDP in the long-run and that causality runs both ways. Thus, a nexus between financial sector development and economic growth exists in Rwanda.

The rest of the paper is organized as follows: section Rwanda's Institutional Environment and Macroeconomic Performance discusses the institutional environment and macroeconomic performance in Rwanda while literature is reviewed in section Overview of Theoretical and Empirical Literature section. The econometric methodology is presented in section Data and Methodology and estimation results and conclusions are drawn in sections Empirical Results and Discussion and Conclusions and Policy Implications respectively.

2 Rwanda's Institutional Environment and Macroeconomic Performance

The Government of Rwanda (GOR) undertook deep economic reforms which allowed the country to shift gradually from a regulated to a market based economy. From a monetary and exchange rate policy viewpoint, direct control measures were gradually relaxed and these gave way to mechanisms and innovations that relied more extensively on market forces to regulate the economy (Rutayisire 2008). Key innovations in the financial markets included the liberalization of current account operations and exchange rate markets, establishing foreign exchange bureaus and authorization of direct investments and transfers of income generated by such investments abroad. In terms of balance of payments, the current account balance as a percentage of GDP has been negative; this is explained by the country mostly experiencing high import bills and low exports.

GDP growth has been gradual. The economy saw a dramatic increase in GDP, peaking at 11 % in 2008 and it has since remained at this high level at least since 2010. Commentators have attributed this robust and sustained GDP growth to the changes that occurred in the institutional environment following various structural reforms and economic stabilization programs, and the international community's active support to the implementation of ongoing programs in EDPRS I and 2. According to the International Monetary Fund (2012), the success can be attributed to better coordination of monetary and fiscal policies supplemented by a prudent monetary policy. Inflation declined from 22 % in 2008 to 2.1 % in 2014. Prudent management of both macro and microeconomic policies helped in building investors' confidence and guiding their decisions for resource allocations better.

Nonetheless, the country faces challenges, some of which are due to its geographical positioning while others are exogenous macroeconomic influences. These include high imports of capital and intermediate goods and energy. This has led to an increase in demand for foreign exchange resulting in occasional depreciation of the Rwandan franc which hit about 1.37 % against the US$ in 2012. This also led to

Table 1 Major indicators of the national economy (%)

Description	2005	2006	2007	2008	2009	2010	2011	2012	2013	2014
Real GDP growth rate (%)	9.4	9.2	7.6	11.2	6.2	7.3	7.8	8.8	4.7	7.0
GDP per capita (USD)	295.0	342.0	409.0	506.0	547.0	572.0	627.0	689.0	701.0	718.0
Investment/GDP (%)	20.3	19.8	22.4	24.6	24.9	24.6	23.7	24.9	24.5	24.2
M3/GDP (%)	17.1	18.7	20.6	17.8	17.5	18.5	20.3	20.1	21.1	22.7
PSC/GDP (%)	11.5	12.3	12.6	13.3	12.1	11.9	13.5	16.2	16.0	17.1
Openness (%)	36.7	37.6	40.3	40.6	38.7	40.1	44.0	46.1	46.3	45.5
Inflation rate (%)	5.6	12.1	6.6	22.3	5.7	0.2	8.3	3.9	3.7	2.1

Source: World Bank Development Indicators (2013), NBR annual reports, various

deterioration in the trade balance. In order to mitigate and deal with these challenges BNR has had to implement flexible and timely policies to smoothen the volatility in market driven exchange rates. BNR maintains a flexible exchange rate regime but undertakes occasional interventions to smoothen exchange volatility.

On the monetary development side, there has been significant improvement in the level of monetization of Rwanda's economy. As shown in Table 1, the ratio of broad money (M3) to GDP rose from 17.1 % in 2004–2005 to 23 % in 2013–2014. Over the same period, the ratio of private sector credit to GDP rose from 3.5 to 14 %, an approximately fourfold increase in real terms (that is, after adjusting for inflation) over the past 10 years.

In terms of level of national competiveness and doing business, Rwanda has been unparalleled both in the sub-region and in Sub-Saharan Africa (SSA). According to the World Bank, Rwanda is the 3rd easiest country to do business in in SSA, two places behind the first, Mauritius and just next to South Africa in the rankings. In terms of competitiveness, the World Global Competitiveness Forum Report (2012–2013) ranked Rwanda 63rd out of 144 countries, leaving behind Kenya (106th), Uganda (123rd), Tanzania (121st) and Burundi (144th). It was the 3rd in SSA and topped the East African Community (EAC). This means that Rwanda has been able to get high scores on each of the three main factors that drive competitiveness: factor driven (institutional, infrastructure, macroeconomic, health and primary education); efficiency enhancers (higher education and training, goods market efficiency, labor market efficiency, financial market efficiency, technological readiness and market size); and innovation driven factors (business sophistication and innovation). This level of competitiveness has been attributed to the government's pursuit of positive macro and microeconomic strategic policy frameworks which are geared at enabling the environment for the development of sound financial systems and spurring robust socioeconomic and sustainable development (EDPRS 2, 2013–2018). These policies have built investor confidence and better guided their decisions of allocating resources (NBR, 2012).

Rwanda has made good progress over the last two decades since it faced enormous challenges in the aftermath of a genocide that destroyed the entire social and economic fabric of the country. It has also benefited from rapid economic growth, reduced poverty, more equality and increased access to services including finance, health and education.[1] In the 10 years to 2010–2011, poverty levels fell from 56.7 % in 2000–2001 to 44.9 %, a reduction of 11.8 % in the population living under the poverty line. The authorities are targeting poverty levels of less than 30 % by 2018.

3 Overview of Theoretical and Empirical Literature

3.1 Theoretical Literature

The causality effect between financial development and economic growth has been a controversial issue for several years. Some researchers have found a positive impact of financial development on economic growth. In cross-country or geographical regions and income groups others have found a significant relationship for some geographical regions and none in others, especially for developing countries. Even though the link/relationship between financial development and economic growth is accepted, the direction of causality is still under debate. In this section we review theoretical and empirical literature that underlies this relationship. There are two key hypotheses that explain the theoretical link between financial development and economic growth—supply and demand leading hypotheses. A supply leading hypothesis has received great support as it has been holding for most countries.

McKinnon (1973) and Shaw (1973) have developed a robust model of financial development appropriate to LDCs, through which financial development positively affects economic growth. Known as a complementarity hypothesis, the McKinnon (1973) and Shaw (1973) model is based on the positive relationship between the real deposit rate of interest and investment; this is contrary to previous thought where this link was negative. The model stresses the negative effects of financial repression on economic growth which characterize developing economies. In fact, they conclude that financial repression through interest rate ceilings, directed credit, exchange rate controls, control on the finance sources of banking institutions and other forms of financial repression are only a recipe for the negative real deposit rate of interest. This reduces the supply of loanable funds and forces banking institutions to apply credit rationing in front of excess demand for loanable funds. The outcome is the allocation of funds not based on the productivity of investments but on other factors like transaction costs and apparent default risks. As a result, there is a tendency to allocate credit to non-productive investments which decrease investment productivity and efficiency, thus slowing down economic growth.

Financial liberalization was proposed as a model of financial development which would lead to economic growth through an increase in the real deposit rate of

[1] EDPRS 2 (2013–2018).

interest, increasing saving mobilization and the financing of the economy both from internal and external sources as a result of capital liberalization. This model has been a central point in analysing the effects of financial development on economic growth where most studies compare before and post financial liberalization periods as shown in Jankee (2006).

The supply leading hypothesis has not received unanimity among economists. Some influential economists such as Robinson (1952) and Friedman and Schwartz (1963) have argued that the development of the financial sector is induced by economic growth such that it comes as a result of higher demand for financial services. Robinson maintains that economic growth creates supply for financial services which will lead to financial development. Levine (2001) argues that economic growth may reduce fixed costs of joining financial intermediaries which enables more people to join, hence financial sector may be caused by improvement in economic growth. Kuznets (1955) supports this view, arguing that finance does not exert a significant impact on economic growth but rather when the economy grows, more financial institutions and financial products (financial innovation) and services come into the market in response to a higher demand for financial services. For Thangavelu (2004), enterprise guides and then finance follows.

3.2 *Empirical Literature Review on the Link Between Financial Development and Economic Growth*

Evidence on the link between financial development and economic growth covers a variety of studies using time series analyses, cross-country growth regressions and panel studies. Among others, Al Nasser and Gomez (2009) establish that financial development is important in GDP growth because it affects investments and the overall cost structure of operations in a country. Kinda (2010) observes that financial development is an engine of economic growth, providing better business opportunities for customers and firms.

Consistent with King and Levine (1993), Levine and Zervos (1996) show that a higher level of financial development is positively correlated to economic development. Their findings suggest that the legal environment facing banks can have a significant impact on economic growth through its effect on bank behavior. Douglas (2003) who investigated evidence of the finance growth nexus in a sample of emerging SSA countries using co-integration and a vector error-correction model found that financial development and economic growth were linked in the long-run in seven of the eight countries and the causality test revealed unidirectional causality from finance to growth in Ghana, Nigeria, Senegal, South Africa, Togo and Zambia. For Ivory Coast and Kenya, the causality ran from growth to finance, confirming the demand leading hypothesis in these two countries.

Spears (1992) examined the causal relationship between financial intermediation and economic growth in a sample of five Sub-Saharan African countries (Burkina Faso, Cameroon, Ivory Coast, Kenya and Malawi), using Granger causality and auto-regressive distributed-lag regressions. He used two measures of financial development: the ratio of money supply to real per capita GDP and the ratio of quasi-money to money supply. He found no causality between the latter and economic growth and these results may be attributable to the wrong measure used rather than to the absence of causality between financial development and economic growth. But the causality from financial development to economic growth was found when the ratio of money supply to GDP was used.

Teame (2002) used a VECM in 13 SSA countries and found from co-integration analyses that there existed a long-run relationship between financial development and economic growth in 12 out of the 13 countries and the causality ran from finance to growth in eight of the countries taken in the sample. Six countries provided evidence of bi-directional causality.

As can be seen, most empirical studies are based on cross-country and panel data analyses. However, countries are heterogeneous and have different experiences in relation to financial development (indicators) and economic growth (Abu-Bader and Abu-Qarn 2008; Greenidge and Moore 2008). Thus, one major limitation of focusing on cross-country regressions is that country-specific questions regarding the influence of the financial sector on the real economy are omitted so that country-based evidence provides the only reliable backdrop against which to assess such linkages.

This paper contributes to financial development and economic growth literature by investigating the link between a number of relevant financial indicators and economic growth, but in one country, Rwanda, over the period 2006Q1–2014Q4. The choice of the study period reflects data availability. A triangulation of VECM and Granger non-causality approaches and impulse response functions were employed to investigate the dynamic inter-relationship between various measures of financial sector development and economic growth and for addressing questions of direction of causality between financial sector development and growth in Rwanda. A main novelty of this paper in the context of financial sector development-economic growth nexus literature is in the use of a rich dynamic approach that allows short-run adjustments and long-run equilibrium relationships to differ.

4 Data and Methodology

4.1 The Econometric Model

4.1.1 Vector Auto-regressive Framework

Based on the Johansen (1988) approach, vector auto-regressive (VAR) methods have become the 'tool of choice' for estimating and testing multivariate relationships among non-stationary data in much of the time series macro-econometrics. Accordingly, in the current application, we allow for rich dynamics in the way GDP adjusts to financial variables and vice versa by nesting the empirical analysis in a VAR framework. In its unrestricted error correction representation, VAR is of the form:

$$\Delta x_t = \Pi x_{t-1} + \sum_{i=1}^{k-1} \Gamma_1 \Delta x_{t-i} + \Phi d_t + \varepsilon_t \tag{1}$$

where x_t is a vector of endogenous variables, $\Gamma_i = (I - A_1 - A_2 - \ldots - A_i)$ and $\Pi = (I - A_1 - A_2 - \ldots - A_k)$ comprises coefficients to be estimated by Johansen's (1988) maximum likelihood procedure using a $(t = 1, \ldots, T)$ sample of data. $i = 1, \ldots, k - 1$ is the number of lags included in the system, Δ is a first difference operator, d_t is a $(m\ x\ 1)$ vector of m deterministic terms (constants and linear trends), Φ is a $(p\ x\ p)$ matrix of coefficients, and $\varepsilon_t \sim \text{iidN}(0, \Sigma)$ is a $(p\ x\ 1)$ vector of errors with zero mean, that is, $E(\varepsilon_t) = 0$, a time-invariant positive definite covariance matrix Σ, and are serially uncorrelated, that is, $E\left(\varepsilon_t \varepsilon_{t-k}^{'}\right) = 0$ for $k \neq 0$.

If at least some of the variables in x_t are unit-root non-stationary then Π in Eq. (1) has reduced rank, which can be formulated as the hypothesis of co-integration:

$$\Pi = \alpha\beta^{'} \tag{2}$$

where α and β are $n\ x\ r$ coefficient matrices and r is the rank of Π corresponding to the number of linearly independent relationships among the variables in x_t. The advantage of this parameterization is in the interpretation of the coefficients. The effect of levels is isolated in the matrix $\alpha\beta$' while Γ_i describes the short-run dynamics of the process, delivering a neat economic interpretation to the vector error correction model of Eq. (1). The r columns of β represents the co-integrating vectors that quantify the 'long-run' (or equilibrium) relationships among the variables in the system and the r columns of error correction coefficients of α load deviations from equilibrium into Δx_t for correction, denoting the speed of adjustment from disequilibrium to ensure equilibrium is maintained. Finding the existence of co-integration is the same as finding the rank (r) of the Π matrix. If it has full rank, $r = n$ and there are n co-integrating relationships, that is, all variables are potentially I(0).

4.1.2 Granger Non-causality Test/Block Exogeneity Tests

Determining in this model that the variables are co-integrated implies there must be Granger causality in at least one direction. On this account, and following Granger (1969), y is said to be Granger-caused by x if x helps in the prediction of y or if the coefficients on the lagged x's are statistically significant in y and vice versa. Thus, the Granger-causality model is specified as:

$$\Delta y_t = \vartheta_1 + \sum_{i=1}^{n} \alpha_{11} \Delta y_{t-n} + \sum_{i=1}^{n} \beta_{12} \Delta x_{t-n} + \varepsilon_{1t} \tag{3}$$

$$\Delta x_t = \vartheta_2 + \sum_{i=1}^{n} \alpha_{21} \Delta y_{t-n} + \sum_{i=1}^{n} \beta_{22} \Delta x_{t-n} + \varepsilon_{2t} \tag{4}$$

where n is the maximum lag-length and ε_{1t} and ε_{2t} are additive stochastic error terms, which are by assumption normally distributed with a zero mean and a constant variance. In light of Eqs. (3) and (4), we can deduce two (2) testable hypotheses (H1 & H2):

(i) that $\sum \beta_{12} = 0$ while $\sum \alpha_{11} \neq 0$, that is, x does not Granger-cause y (no causality from x to y), treating x as either a variable or a vector of variables.
(ii) that $\sum \beta_{22} \neq 0$ while $\sum \alpha_{21} = 0$, that is, y does not Granger-cause x (no causality from y to x), treating y as either a variable or a vector of variables.

Acceptance of either hypothesis will suggest the existence of unidirectional causality between x and y, and feedback between x and y may be understood to exist if $\sum \beta_{i2} \neq 0$ and $\sum \alpha_{i1} \neq 0$. Alternatively, no causality between x and y exists if $\sum \beta_{i2} = 0$ and $\sum \alpha_{i1} = 0$.

The first step in estimating this model is the choice of variables that should be included in x_t in Eq. (1). Four variables used in this paper for economic specifications include an indicator of output growth, that is, GDP_t, and a host of financial indicators namely the $M3_t$, $tb91_t$ and the exr_t over which monetary authorities in Rwanda exercise control both directly and/or indirectly. GDP is measured in billions of Rwandan francs at constant 2011 prices. Money supply ($M3_t$) which includes currency outside the central bank plus demand, savings and time deposits and foreign currency deposits, is used as a measure of broad money. Besides, it is the target that Rwandan monetary authorities use in influencing the level of economic activities in the economy, nominal exchange rates (exr_t) used is the average rate of FRW/US dollar, and the 91 day treasury bill rate ($tb91_t$) is a proxy for the National Bank of Rwanda (NBR) policy rate. The link between tb91 and the NBR policy rate is the first stage in the interest rate transmission mechanism. Moreover, data on the NBR policy rate is only available from the date it started, that is, January 2008, and if used could reduce further the already small sample. The subscript t reflects the time series nature of the data and is measured on a quarterly frequency.

4.2 Variables, Data Measurement and Transformations

Quarterly time series data for 2006Q1–2014Q4 was used for the four variables investigated. Much as different parameters can be used to measure financial development and economic growth, we considered three financial indicators which are assumed to capture a strong influence on economic growth. In this paper, the financial sector parameters used are: broad money supply ($M3_t$), nominal exchange rates (exr_t), and the NBR policy rate (pcr_t). The level of economic activity was measured using GDP_t and was measured in billions of Rwandan francs at constant 2011 prices. Due to low levels and the agricultural nature of income and predominance of a transaction motive as well as having limited or negligible financial assets, *M3*, which includes currency outside the central bank plus demand, savings and time deposits and foreign currency deposits, was used as a measure of broad money. Besides, it is the target that Rwandan monetary authorities use in influencing the level of economic activity in the economy.

Some researchers have adopted monetary aggregate M1 (the currency in circulation held by the public plus demand deposits in banks). Others prefer a broader measure of money supply, M2. In this study we used M3 which appears more stable in relation to money income but less controllable (Rutayisire 2008). Moreover as indicated earlier, this aggregate is used by NBR as the intermediate target to conduct monetary policy. pcr_t is used to determine sensitivity of interest rate to demand for money growth and ultimately in real GDP. In particular, the inclusion of pcr_t is premised on the fact that changes in it are expected to influence longer term interest rates, notably time deposit and bank lending rates which are deemed to be fundamental for monetary policy transmission to prices and the real economy.

The nominal exchange rate used was the average rate of Rwandan franc (Rwf) as per the US$. Change in the US$/Rwf exchange rate was used as a proxy for the anticipated depreciation of the domestic currency and was calculated from the official exchange rate series. Depreciation of the domestic currency leads to a reduction in demand for money balances in the domestic currency in favor of the foreign currency. Being a net importer, Rwf has been depreciating against the other currencies in EAC countries resulting in higher demand for foreign exchange that pushes the depreciating Rwf. All data on the five endogenous variables used in the analysis were from the NBR database, and were, save for pcr_t expressed in logarithms and are shown in levels and also in first differences in Fig. 1.

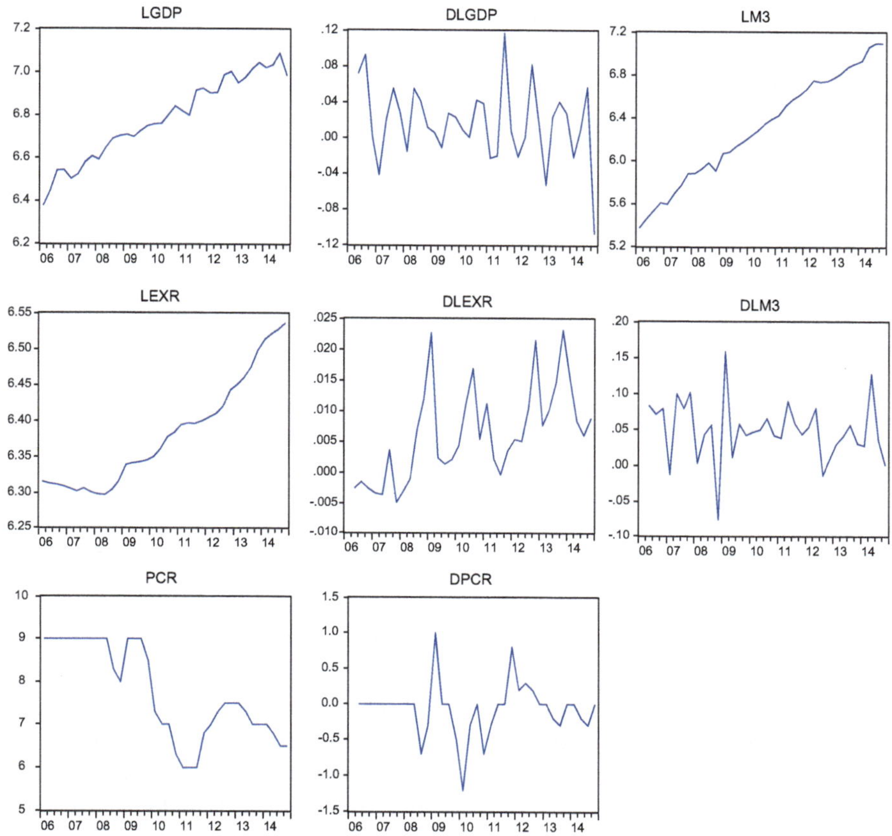

Source: Authors' computation.

Fig. 1 Series plots in level and first differences. *Source*: Authors' computations

5 Empirical Results and Discussion

As a precursor to a co-integration analysis, it is customary to begin with the graphical expositions of the level and first difference of the series to reveal important data features. Visual inspection of the data in Fig. 1 shows that all variables typically trend over time or follow the same pattern, that is, they are not stationary as they are not mean-reverting in levels. However, in first differences, they seem to be mean-reverting and could potentially be stationary. After a visual inspection, the series were tested for the order of integration or non-stationarity using the Augmented Dickey Fuller (ADF) unit root test (Dickey and Fuller 1979). Mindful of the fact that critical values of the t-statistic depend on whether an intercept and/or time trend is included in the regression equation and on the sample size (Enders 2010: 206), the τ_τ—statistic, scaled by the 5 % critical value for $n = 50$ usable observations were used. The statistic critical values were obtained from Table A in Enders (2010: 488).

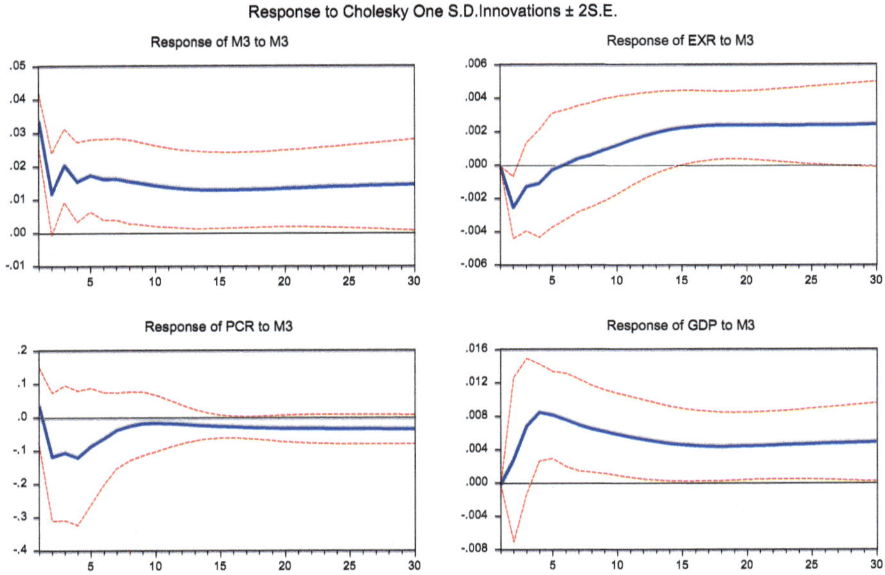

Source: Authors' computations.

Fig. 2 Responses to money supply shocks. *Source*: Authors' computation

The results of unit root test are given in Table 2, and as expected they indicate that all variables are non-stationary, but become stationary upon first differencing at the conventional 5 % level of significance. Furthermore, because our data was high frequency (that is, quarterly observations), all series, save for the nominal exchange rate, were adjusted for seasonal effects, although this was only marginally significant in the GDP series.

On the basis of unit root testing, we treated EXR_t, GDP_t, $M3_t$ and PCR_t as unit root non-stationary and the fact that they are integrated of the same order, that is, I(1) could be co-integrated. The unrestricted 4-dimensional model was estimated with a restricted constant. The choice of the lag-length was determined as the minimum number of lags that meet the crucial assumption of time independence of the residuals, based on a Lagrange Multiplier (LM) test. We began with $k = 5$ lags. Both Schwarz Bayesian criterion [SC] and Hannan-Quinn Criterion [HQ] supported one lag while AIC favored two lags. Fortunately, with either $k = 1$ (as chosen by SC and HQ) or $k = 2$ (as chosen by the AIC), the LM test could not reject the null hypothesis of no serial correlation in the residuals.[2] So we chose to estimate the underlying model using two lags to capture much more rich dynamics in the system.

An assessment of the VAR (2) system residual mis-specification test revealed that although the residuals met the crucial assumption of no auto-correlation and the

[2] LM test for VAR (1) and VAR (2) residuals were [$\chi^2(16) = 16.392[0.426]$] and [$\chi^2(16) = 13.394$ [0. 644]], respectively.

Table 2 The Augmented Dickey-Fuller (ADF) unit root test

The Augmented Dickey-Fuller (ADF) unit root test					
	ADF test in level			ADF test in first difference	
Macrovariables	$H_0 : \gamma = 0$	Lag-length	Inference	$H_0 : \gamma = 0$	Inference
$LEXR_t$	−1.659 (−3.563)	4	I(1)	−3.825 (−3.563)	I(0)
$LGDP_t$	2.324 (−1.951)	0	I(1)	−4.992 (−1.951)	I(0)
$LM3_t$	−2.456 (−3.552)	2	I(1)	−8.194 (−2.951)	I(0)
PCR_t	−1.009 (−1.951)	1	I(1)	−4.237 (−1.951)	I(0)

Notes: L = log; EXR = bilateral nominal exchange rate (FRW/US); GDP = 2011 constant price gross domestic product, M3 = broad money; PCR = NBR Policy rate; and $\gamma = 0$ is the null hypothesis that the series contains a unit root. Modified Akaike Information criterion [AIC] was used (maximum set at 9 lags) and the appropriate lag-length determined through the very painful general to specific procedure. ADF equations for GDP and PCR were estimated with no intercept and trend (or none option in E-views) because these were found insignificant. For the same reason, the first difference estimation of the ADF equation for M3 included only the intercept (intercept option in E-views). Elsewhere, an unrestricted intercept and restricted linear trend were included in the ADF equations in both levels and first difference. Numbers in parenthesis are the 5 % critical values, unless otherwise stated. All unit-root non-stationary variables are stationary in first differences

Source: Authors' computations

white Heteroskedasticity test was satisfactory [$\chi^2(160) = 181.61[0.116]$], the assumption of multivariate normality of the residuals was not met [$\chi^2(8) = 92.521$ $[0.000]$]. Non-normality notwithstanding, the good news is that estimates of the VAR model are robust to deviations from normality provided residuals are not auto-correlated (Bwire et al. 2013b:10; Bwire 2012), as indeed is the case in the system analysed here.

Having determined the appropriate specification of the data generating process, the existence of long-run equilibrium relation (s) was then determined using the Johansen (1988) trace statistic test for co-integration.[3] Based on the results Table 3, the presence of one equilibrium (stationary) relation among the variables at the conventional 5 % level of significance cannot be rejected.

5.1 *Vector Error Correction Model (VECM) Analysis*

Unless otherwise noted, the only existing co-integrating relation, as shown in Table 3, is normalized for the quarterly change of GDP in order to interpret the estimated coefficients. Results of the long-run analysis reveal that broad money (M3) has a positive significant long-run correlation with GDP while GDP's

[3] In the test, the determination of the co-integrating rank, r relies on a top-to-bottom sequential procedure. This is asymptotically more correct than the bottom-to-top alternative (that is the Max-Eigen statistic) (Juselius 2006: 131–134).

Table 3 Johansen's co-integration test and long-run analysis

Hypothesized no. of CE(s)	Eigen value	Trace statistic	0.05 critical value
None*	0.662723	74.25490	63.87610
At most 1	0.421851	38.38885	42.91525
At most 2	0.366720	20.30735	25.87211
At most 3	0.146604	5.231551	12.51798
Normalized cointegrating equation:			
LGDP	LEXR	LM3	PCR
β' 1.000	0.067 (0.572)	0.312 (14.141)	−0.008 (−1.832)
α 1.198 (3.056)	−0.184 (−2.564)	0.678 (1.286)	−0.719 (−0.126)

Notes: Trace test indicates 1 cointegrating eqn(s) at the 0.05 level; Constant/Trend assumption: Restricted constant; * denotes rejection of the hypothesis at the 0.05 level; The model uses 2 lags, and in parentheses are *t*-values
Source: Authors computation

correlation with the policy rate is negative. The long-run estimated coefficient to M3 is about 0.3, indicating that holding other factors constant, a 1 percentage point increase in M3 should lead to a long-term increase in GDP by about 0.3 percentage points. While a reduction in credit to the private sector by 1 percentage point hurts GDP growth by 0.01 percentage points in the long-run, albeit boundary significant so., holding everything else constant.

VECM estimates revealed that the error correction term was statistically significant with a correct sign for the exchange rate equation, adjusting by about 18 % of the previous quarter's deviation from equilibrium, suggesting that the exchange rate influences GDP over the short-run period.

5.2 *The Engle Granger Non-causality Test*

Turning to the direction of causality, two elements are of interest: (i) have measures of financial sector development or concentration collectively influenced GDP growth in Rwanda so that causation runs from financial sector development to growth?, (ii) is it the robust and sustained GDP growth over the last two decades that Rwanda has achieved that has been an impetus to its financial sector development? If it happens to be the case that we fail to reject both (i) and (ii) simultaneously, it may well be the case that a nexus exists in Rwanda where financial sector development is key to accelerating the desired rate of growth and at the same time that growth is a stimulus to financial sector development. The results of both pairwise and block exogeneity are given in Table 4.

Collectively, the null hypothesis that a host of financial indicators over which the monetary authorities in Rwanda exercise control both directly and/or indirectly do not influence growth in the short-term is rejected. Therefore, developments in the financial sector lead to growth. Individually, however, money supply, a key target variable for Rwandan monetary authorities does not support growth in the short-

Table 4 Granger non-causality tests/block exogeneity wald tests

	Chi-sq	d.f	Prob.	Chi-sq	d.f	Prob.	Chi-sq	d.f	Prob.	Chi-sq	d.f	Prob.
GDP	–			2.029	2	0.363	0.440	2	0.802	0.702	2	0.704
EXR	2.888	2	0.236	–			4.761	2	0.093	0.432	2	0.806
M3	5.440	2	0.066	7.426	2	0.024	–			7.098	2	0.02
PCR	0.458	2	0.795	3.323	2	0.189	4.746	2	0.093	–		
All	7.049	6	0.136	26.150	6	0.000	8.951	6	0.176	10.254	6	0.114

Note: "–" denotes dependent variable
Source: Author's computations

term. This is not surprising because monetary policy has a lagged/delayed impact on the real economy as indeed, money supply in the long-run significantly supports growth. Moreover, in the short-to medium term, the link between M3 and GDP is strongly weakened by the prohibitively high lending rates.

As is the case elsewhere in the region, the transmission mechanism of monetary policy to real rates remains asymmetric in Rwanda. NBR has been less successful so far in influencing bank lending rates which are often sticky downwards even if the NBR policy rate is on a downward trajectory. Estimates by staff members of NBR Research Department and IMF indicate that the bank lending rate responds to a 100 basis point change in the NBR policy rate by about 3 and 2 basis points in the short and long term, respectively. It is, however, important to note that the high lending rates are not only unique to Rwanda, but are a feature of most banking systems in Africa with pretty similar causes and reflect the high spreads between bank deposit and lending rates. Rwandan banks have the third highest spreads in EAC, after Burundi and Uganda. The largest single contributor to interest rate spreads in Rwanda is bank overhead costs. Overheads are high because of the structural features of the economy and the banking system. In particular, the financial systems are still quite shallow, which implies that banks are not achieving economies of scale which would allow them to operate more efficiently and hence reduce their overheads. Banking activities are also impeded by a poor institutional environment, characterized by a slow and inefficient legal system, unreliable valuation of properties for collateral purposes and difficulty in selling property.

Similarly, the null hypothesis that GDP does not influence financial variables in the short-run is rejected for all block equations of financial indicators. Therefore, the aggregate level of demand in the economy has a direct influence on developments in the financial sector, that is, GDP growth leads financial sector development. As causation runs both ways (from financial development indicators to GDP and from GDP to financial development indicators), a nexus exists between growth and financial sector development in Rwanda. This result is consistent with Levine (2001), Thangavelu (2004) and Bwire and Musiime (2008) for Uganda.

Overall, our empirical results for the sample period analysed suggest that money supply, a key target variable for Rwandan monetary authorities does not seem to necessarily trigger the level of economic activities in the short-run, but it does so in the long-run as a 1 percentage point increase in M3 induces about 0.3 percentage points change in GDP. Thus, short-term monetary policy interventions geared at supporting growth in Rwanda need not focus on changes in money supply, but rather on other monetary policy tools, in particular the policy rate which is synonymous with the inflation targeting framework.

5.3 *Impulse Response*

The downside to both long-run and short-run estimates is that they are partial derivatives (by construction) predicated on the ceteris paribus clause (Lütkepohl and Reimers 1992), and have in this paper been interpreted in this light. As argued by Lloyd et al. (2006) and Bwire et al. (2013a), where variables in an economic system are characterized by potentially rich dynamic interactions (as is the case here), inferences based on 'everything else held constant' are both of limited value and may give a misleading impression of the short and long-run estimates. Therefore, since what we want is to actually estimate what might happen to all variables in the system following a perturbation of the known size in the money supply[4] equation, an impulse response analysis, which describes the resulting chain reaction of knock-on and feedback effects as it permeates through the system, provides a tractable and potentially attractive value of the transmission mechanism of money supply provided no other shocks hit the system thereafter.

The results of money supply shock are shown in Fig. 2. Specifically, Fig. 2 shows the impact of a one standard deviation shock, defined as an exogenous, unexpected, temporary increase in money supply with a 95 % confidence level on the nominal exchange rate, policy rate and GDP in period zero. The solid line in each graph is the estimated response while the dashed lines denote a two standard error confidence band around the estimate. Since the data are in first differences of logarithms, save for the policy rate, the impulse response functions need to be regarded as measuring a proportional change in the rest of the macro-variables due to one standard innovation (at the initial period) in money supply.

It is clear from Fig. 2 that the effect of money supply shock on GDP is fairly gradual (taking about four quarters to reach the full impact). The immediate effect of a structural one standard deviation shock which is equivalent to a 0.034 percentage points reduction in money supply leaves output growth unchanged and stable, with an impact elasticity of 0. Consistent with the Granger non-causality test, this suggests no short-term transmission of the monetary aggregate, M3, on the real economy and is due to the high interest rate spreads discussed earlier. However, the full effect of this shock is realized after about four quarters, and is about a 0.009

[4] This is because as mentioned earlier, it is a key target variable for the Rwandan monetary authority.

percentage point increase in GDP, implying a dynamic elasticity of monetary aggregates' transmission mechanism of 0.25.

As shown in Fig. 2, money supply shocks, specifically a reduction in money supply triggers appreciation pressures in the exchange rate which eases after reaching a full effect in two quarters, but depreciates thereafter. With respect to the central bank's (NBR) reaction, the figure indicates that a temporary appreciation in the exchange rate is followed by a monetary policy loosening especially as the appreciated exchange rate does not pose a threat to macroeconomic stability. Thereafter the central bank tightens its reaction (probably informed by a depreciating exchange rate over the same period), and eventually settles on its long-run equilibrium path.

Importantly, the impulses show that monetary policy actions in Rwanda are reactionary as the rise in the policy rate is delayed compared to the overheating in the economy, that is, while GDP peaks in the fourth quarter following an initial shock in money supply, the policy rate responds to this overheating only in the fifth quarter. This suggests that to anchor expectations of economic agents and to maintain monetary policy credibility, a forward looking monetary policy framework might be the way to go.

6 Conclusions and Policy Implications

This paper investigated the role of financial sector development in stimulating growth and, at the same time the notion that economic growth stimulates financial development using data for 2006Q1–2014Q4. It used a triangulation of VECM and Granger non-causality approaches and impulse response functions. Our empirical analysis yields evidence that is strongly supportive of a positive influence of financial development on economic growth, but in the long-run. This finding is consistent not only with existing empirical literature, most of which uncovers a positive relationship between finance and growth, but also with economic theory.

Economic theory posits a number of intuitively plausible rationales for why well-functioning financial systems matter for growth. These include producing information and allocating capital; monitoring firm behavior and exerting corporate governance; facilitating the hedging, trading and pooling of risks; mobilizing savings for investment; and reducing transaction costs of economic exchange and activity. An accurate measure of financial development will capture how well commercial banks perform these growth-promoting functions.

There are a number of key messages which emerge for the country's policymakers from this empirical analysis. Above all, our central finding that financial development has a positive, significant effect on growth, suggests that financial development will be a key ingredient for the country's medium and long-run growth. The obvious corollary is that policy efforts seeking to further strengthen and improve Rwanda's financial system will yield dividends for growth going forward. In addition, our evidence indicates that what matters more for developing

Rwanda is overall financial development rather than the development of particular components of the financial system.

Our results also show that monetary policy continues to be supportive of economic growth in Rwanda as the policy rate responds to overheating of the economy which suggests that the policy rate plays a key signalling role for the financial system and the economy at large. However, the fact that the reaction occurs five periods after the shock suggests the need for adopting a forward looking monetary policy framework to appropriately anchor expectations of economic agents and for maintaining monetary policy credibility. In the very short-run, the level of real economic activity is not particularly sensitive to the key target variable of money supply used by Rwandan monetary authorities.

The main constraint to this very important transmission mechanism in Rwanda is the relatively shallow and undiversified financial system. It is shallow in terms of the quantity of financial assets in relation to GDP and narrow in terms of its lack of diversity. Consequently, policy measures to strengthen the transmission of monetary policy actions to the real economy in the short to medium term should be focused on deepening and broadening the financial system, especially through reforms of those institutions that have a positive influence on both banks and capital markets including the liberalization of the pension sector and reforms to the land registry and the commercial justice systems to reduce credit risks. Maintaining macroeconomic stability is also essential for financial sector development because without macroeconomic stability savers will not hold financial assets.

Acknowledgments The authors gratefully acknowledge assistance from staff of the Research Department at the National Bank of Rwanda, and the useful comments from participants in a conference on 'Recent trends in Economic Development, Finance and Management Research' in the developing economies of East Africa (EABEW), held on 5–6 May 2015 at the University of Rwanda, College of Business and Economics (CBE) Kigali.

References

Abu-Bader S, Abu-Qarn AS (2008) Financial development and economic growth: empirical evidence from six MENA countries. Rev Dev Econ 12(4):803–817

Al Nasser OM, Gomez XG (2009) Do well-functioning financial systems affect the FDI flows to Latin America. Int Res J Finance Econ 29:60–75

Bwire T (2012) Aid, fiscal policy and macroeconomy of Uganda: a cointegrated vector autoregressive (CVAR) approach. PhD Thesis, University of Nottingham, Nottingham, UK

Bwire T, Musiime A (2008) Financial development—economic growth nexus: empirical evidence from Uganda. A paper presented at the 2008 AEC conference, Tunis, Tunisia

Bwire T, Anguyoand FL, Opolot J (2013a) Exchange rate pass-through to domestic prices in Uganda: evidence from a structural vector auto-regression (SVAR). J Stat Economet Meth 2 (4):41–68

Bwire T, Morrissey O, Lloyd T (2013b) A time series analysis of the impact of foreign aid on central government's fiscal budget in Uganda. WIDER Working Paper No. 2013/101

Dickey DA, Fuller WA (1979) Distribution of estimators for autoregressive time series and unit roots. J Am Stat Assoc 74(2):427–431

Douglas K (2003) The finance growth nexus: evidence from Sub-Saharan Africa. IAER 9(2):172
Enders W (2010) Applied time series econometrics. Wiley, New York, NY
Friedman M, Schwartz A (1963) A monetary history of the United States. Princeton University Press, Princeton
Granger CWJ (1969) Investigating causal relationships by econometric models and cross-spectral methods. Econometrica 37:424–438
Greenidge K, Moore A (2008) The impact of financial liberalization on financial development: evidence from the Caribbean. In: Craigwell R, Greenidge K, Williams M (eds) Aspects of financial liberalization and capital market development in the Caribbean. Central Bank of Barbados, Bridgetown, Barbados, pp 111–132
Gurley JE, Shaw ES (1955) Financial aspects of economic development. Am Econ Rev 4:515–538
Jankee CK (2006) Financial globalization and economic resilience in Mauritius. In: Briguglio L, Cordinaand G, Kisanga J (eds) Building the economic resilience of small states. University of Malter and Commonwealth Secretariat, Malter, pp 97–112
Johansen S (1988) Statistical analysis of cointegration vectors. J Econ Dynam Contr 12 (2–3):231–254
Juselius K (2006) The cointegrated VAR model: methodology and applications, advanced texts in econometrics. Oxford University Press, Oxford
Kinda T (2010) Investment climate and FDI in developing countries: firm-level evidence. World Dev 38:498–513
King RG, Levine R (1993) Finance and growth: Schumpeter might be right. Q J Econ 108 (3):717–737
Kuznets S (1955) eCONOMIC growth and income inequality. Am Econ Rev XLV(1):1–30
Levine R (2001) Stock market, growth and tax policy. J Finance 46(4):1445–1465
Levine R, Zervos S (1996) Stock markets, banks and economic growth. The World Bank Policy Research Working Paper, No. 1690
Lloyd T, McCorriston S, Morgan CW, Rayner AJ (2006) Food scares, market power and price transmission: the UK BSE crisis. Eur Rev Agr Econ 33:119–147
Lütkepohl H, Reimers HE (1992) Impulse response analysis of cointegrated systems. J Econ Dynam Contr 16(1):53–78
McKinnon RI (1973) Money and capital in economic development. Brookings Institution, Washington, DC
National Bank of Rwanda (2005) Annual reports. Kigali, Rwanda
National Bank of Rwanda (2009) Annual reports. Kigali, Rwanda
National Bank of Rwanda (2012) Annual report. Kigali, Rwanda
National Bank of Rwanda. Monetary policy and financial stability statement, February 2014, August 2014 and February 2015. Kigali, Rwanda
Patrick HT (1966) Financial development and economic growth in under developed countries. Econ Dev Cult Change 14(1):174–189
Robinson J (1952) The generalization of the general theory: the rate of interest and other essays. Macmillan, London, pp 69–142
Rousseau LP, Sylla R (2001) Financial systems, economic growth, and globalization. NBER Working Paper No.8323
Rutayisire JM (2008) Liberalization, monetary and demand for money in Rwanda. Rwanda. EDPRS 1&2. Kigali, Rwanda
Schumpeter JA (1932) The theory of economic development. Harvard University Press, Cambridge, MA
Shaw ES (1973) Financial deepening in economic development. Oxford University Press, Oxford
Spears A (1992) The role of financial intermediation on economic growth in SSA. J Dev Stud 13:361–380
Teame G (2002) Financial development and economic growth in Sub-Saharan African countries: evidence from time series analysis. Morehead State University, Morehead, KY

Thangavelu SM (2004) Financial development and economic growth in Australia: an empirical analysis. Empir Econ 29(2):247–260

The World Bank. Retrieved from World Development indicators Database: http://web.worldbank.org

Effects of a Common Currency on Intra-regional Trade in Africa: Perspectives on the East African Community Monetary Union

Yvonne Umulisa

Abstract This paper discusses two aspects about a monetary union: first, it presents theoretical arguments and empirical findings suggesting that overall a monetary union itself positively affects intra-union trade which in turn leads to tightly correlated business cycles across countries in the union. Second, using the gravity equation it presents results from both Tobit and OLS models that show the main determinants of bilateral trade among African countries. While common currency does not seem to significantly affect bilateral trade, all traditional gravity model variables including GDP, distance, common border, common language and free trade agreements (FTAs) show the expected sign and high significance. Hence, the paper suggests that these are the main determinants of bilateral trade in African countries and within East African Community (EAC) partner states as well. The paper also shows that pairs of countries joining any FTA, trade about two times as much with each other than they do with countries not in the FTA. Countries using CFA trade more compared to countries that do not use that currency—the effect was between two and four times in 2000 and 2012 respectively, albeit Tobit's results show unexpected signs. In addition the paper also finds evidence of trade effects within EAC countries. In light of these results, the need for an East African Community Monetary Union can be supported and its on-going process should remain the primary target for partner states.

Keywords Optimum currency area • Endogeneity • Gravity model • Bilateral trade • Monetary union

Y. Umulisa (✉)
Jönköping International Business School, Jönköping, Sweden

College of Business and Economics, University of Rwanda, Kigali, Rwanda
e-mail: Yvonne.Umulisa@ju.se

A. Heshmati (ed.), *Economic Integration, Currency Union, and Sustainable and Inclusive Growth in East Africa*, Advances in African Economic, Social and Political Development, DOI 10.1007/978-3-319-30432-8_5

1 Introduction

The introduction of a common currency is without any doubt one of the biggest achievements of partner states in the process of furthering economic integration. That is why, we observe a number of different existing monetary unions (MUs) and potential monetary union processes being considered and developed on every continent (Horvath and Komarek 2002).

Until the surfacing of the sovereign debt crisis within the European Union (EU) countries, the best known and economically strongest MU was the European Monetary Union (EMU) commonly known as the Euro Zone. However, the famous sovereign debt crisis has led many observers to re-evaluate EMU. That is, to try to answer the question whether the Euro Zone can become an optimum currency area (OCA) that its creators initially believed it to be (De Grauwe 2011). Since the establishment of a MU and its membership are irreversible, it is important to be fully aware of the possible problems related to the membership of a MU and of the coordination of policy among its members to reduce risks of failure and the cost of crises. The Euro Zone's current experience should be a warning for different economic integrations, especially EAC, that are envisaging monetary unifications.

African countries have also taken part in a move towards MU; in particular, the EAC recently took steps in the process of its economic integration by envisaging a common currency in the near future (probably in 2024). In November 2013 the heads of partner states signed a protocol on establishing the EAC monetary union, which in Article 3 states that the MU aims at facilitating economic integration to attain sustainable growth and the development of the community (EAC 2013, p 7). The critical questions here are: Are EAC partner states ready to form a MU? Does the EAC form a base for an OCA? How can the experience from EU be used to avoid monetary and macroeconomic policy problems that are currently characterizing EMU?

To answer these questions, different economists use the OCA theory which mainly addresses two issues: the advantages (benefits) and disadvantages (costs) of adopting a common currency, and the characteristics/criteria that are required for countries to form a MU and for the implementation of a common currency. According to Robert Mundell (1961) and many others' vision (Bayoumi and Eichengreen 1997; De Grauwe 1992; Frankel and Rose 1998; Horvath and Komarek 2002; Kouparitsas 2001; Mkenda 2001), for a successful MU, the partner states should mainly fulfill four criteria: higher degree of intra-regional trade, similarity of economic structures (shocks and business cycles), labor mobility and finally a strong system of risk sharing, usually through fiscal transfers. Hence, the greater any of the four criteria, the more the partner states are likely to take advantages of joining a MU.

In view of this, several authors who have studied the costs and benefits to countries intending to form a monetary union by using the OCA theory have not been able to find clear answers, mainly because the OCA criteria taken together are endogenous (Frankel and Rose 1998). With these ambiguities, members of the

Economic Union face a 'critical question' concerning the process and time of implementing a currency union when there is the political and economic will to create one. Hence, over the years, due to developments in macroeconomic theory, the OCA theory has been extended and modified and many of the existing studies use the elements of the 'new theory' of OCA. The authors believe that the OCA test could be satisfied ex-post even if it is not fully satisfied ex-ante; this is the 'Endogeneity of OCA', that is, countries which initially do not fulfill the criteria of an OCA may over time turn into an OCA (Frankel and Rose 1997). The intuition behind the endogeneity hypothesis is that the borders of new currency unions can be drawn larger in expectation that trade integration and income correlation will deepen once a currency union is created therefore facilitating a movement into the OCA (De Grauwe and Mongelli 2005).

Given the political and economic will of creating a MU within EAC member countries, this paper seeks to show to what extent the EAC has the potential of an OCA by exploring the endogeneity hypothesis. Thus, starting with the first two criteria, this paper looks at the existing empirical literature by reviewing the EMU experience to find out whether the Euro Zone's performance has met any of these criteria. In addition, it also focuses on trade flows in African countries to better understand what the main determinants of bilateral trade within them are. Therefore, in light of the findings one may support the EAC MU process.

The remainder of this paper is organized as follows: Literature Review section presents a survey of relevant literature. Data and Methodology section explains the data and methodology used in both descriptive and econometric analyses. The descriptive analysis is discussed in Descriptive Statistics Section. Econometric section discusses the estimation results and findings. The last section provides a conclusion.

2 Literature Review

2.1 The Common Currency Effect on Trade

Important literature exists on the common currency effects on trade in different continents with a large number of works discussing EMU. In what follows, some of the researches focusing on the effects of MU on trade and its repercussions for the symmetry of business cycles in specific cases are provided. A decision by an individual country to join a MU, even if it does not satisfy the OCA criteria can have a self-fulfilling character as currency unions are generally believed to have benefits. Many economists (according to the optimistic view)[1] have argued that the

[1] The optimistic view claims that European trade is dominated by intra-industry trade such that it is based on economies of scale and product differentiation thus leading to similar specialization patterns. As such integration leads to more equal economic structures and less asymmetric shocks.

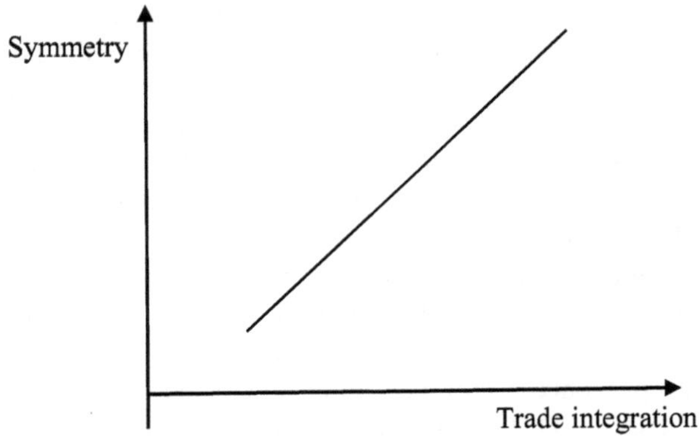

Fig. 1 The Optimistic view

degree of asymmetry among countries decreases as integration (trade) increases but trade integration increases as well when countries join a MU (Rose 1999). As such, the process of integration is speeded up by a very strong decision to join a monetary union, so that the countries' candidates even if they do not belong to OCA will move into the OCA zone once they form a MU and OCA becomes endogenous (Frankel and Rose 1998). The positive relationship between trade integration and the symmetry among countries is given in Fig. 1

Many researchers, including Frankel and Rose (1998), have demonstrated the endogenous effects of MU which favor this optimistic view and which confirmed that monetary unions in general, and EMU in particular, lead to significant increase in trade, and thus trade integration and symmetry evolve over time (Fig. 1).

Rose (1999) analysed the panel dataset including bilateral observations from 186 countries (over 100 pairings and 300 observations in which both countries used the same currency), for 5 years spanning 1970 through 1990. He found a large positive effect of a currency union on international trade and a small negative effect of exchange rate volatility. These effects are statistically significant and imply that two countries that share the same currency, trade three times as much as they would with different currencies. He concluded that currency unions like EMU may thus lead to a large increase in international trade.

Calderon et al. (2007) used annual information for 147 countries for the period 1960–1999 and their findings suggest that differences in the responsiveness of cycle synchronization to trade integration between industrial and developing countries can be explained by differences in the patterns of specialization and bilateral trade. The results reflected the fact that industrialized countries have similar production structures and their intra-trade is dominated by intra-industry trade and thus they increase the synchronization of business cycles among them. In contrast developing countries have different production structures and inter-industry trade.

Recently, Rose (2004) used a meta-analysis technique and by combining empirical results from 34 recent studies that have investigated the effects of currency unions on trade, confirmed a robust, positive effect of currency unions on trade, which remain statistically significant and economically important even after filtering out likely publication selection. Rose concluded that although the combined estimates varied from roughly 30 to 90 %, depending on the exact meta-methods used, they all implied a substantial increase in trade.

Micco et al. (2003), by estimating the early effect of the EMU on trade using panel data from 1992 to 2002 found that the effect of EMU on bilateral trade between member countries ranged between 5 and 10 %, when compared to trade between all other pairs of countries, and between 9 and 20 % when compared to trade among non-EMU countries.

Adam and Chaudhry (2014), studied the currency union (CU) effect on intra-regional trade in the Economic Community of West African States (ECOWAS) using panel dynamic ordinary least square to examine short and long-run currency union effects on intra-ECOWAS trade. They found evidence of a significant positive CU effect on aggregate intra-ECOWAS trade. In addition, their findings revealed that CU is good for aggregate intra-regional trade though some individual members respond in a negative way to CU.

Frankel and Rose (1998), by using 30 years data on 21 industrialized countries found a strong and striking empirical finding: 'countries with closer trade links tend to have more tightly correlated business cycle'. Their findings have led to a number of conclusions on the prospects and desirability of EMU.

Throughout these findings, we can see that the effect of common currency on trade development, which in turn leads to synchronization of business cycles, is an important issue which appears as one of the official motivations behind creating a MU, especially in the creation of EMU.

However, the positive results mentioned earlier have not been supported subsequently. For instance, Baldwin (2006) who reassessed the methodology and principal findings of the pre-EMU literature on the effects of currency unions on trade found that these effects recalibrated, but remained important. For him, after the start of EMU, the euro has already boosted intra-euro area trade by 5–10 % without trade diversions to the rest of the world. He adds that detailed theoretical hypotheses are still needed to better understand what drives these effects and how deep the trade effect can be from the euro for countries in the euro area. Given that trade among European countries has continuously risen over the last 50 years, he concludes that it may be difficult to witness further spectacular surges in intra-European trade, so the full trade effects from EMU may require more time to fully display themselves. In addition, due to the eruption of the sovereign debt crisis which began 6 years ago and is persistent again with the failure of the Greek government to repay its debt, one may say that the Euro Zone has not yet met the endogeneity hypothesis.

2.2 Determinants of Trade

Theoretical and empirical studies have found that trade flows between countries depend on a number of factors including GDP and distance between countries where distance can represent history, culture, language and many other common aspects such as a border (Jafari et al. 2011). Other empirical studies point out the effects of exchange rates on trade flows. In theory it is well know that a change in exchange rate will effect exports and imports. For instance, currency devaluation can improve trade flows if the relative prices among the trading countries and other factors are unchanged. Hence, many international trade researchers confirm that trade flows respond to exchange rates. However, some of them do not find any significant effects of exchange rates (Rose 1999).

Using the described theoretical framework given earlier, our main conclusion is that MU has important benefits, but it also imposes costs. MU increases trade integration which decreases asymmetric shocks that may occur between member MU countries. However, there are other economic aspects of the partner states that can create asymmetric shocks which will need forces stronger than intra-regional trade to adjust. That is why the OCA theory suggests that a certain number of conditions such as price flexibility or labor mobility should be in place for member countries to fully benefit from forming a MU (De Grauwe 2014).

3 Data and Methodology

On the one hand, a descriptive analysis relies on measuring two different indicators trade integration and GDP business cycles for selected countries of EMU and for all EAC countries is used and on the other hand an econometric analysis based on the gravity model of trade—the empirical methodology most commonly used for the determinants of bilateral trade—is also used to better understand factors that drive bilateral trade in Africa. The gravity model, in its basic form, assumes that trade between countries can be compared to the gravitational force between two objects where the force of gravity between the two objects is proportional to the product of the masses of the two objects divided by the square of the distance between them. Hence, in trade theory, the force of gravity is replaced with the value of bilateral trade and the masses with the trade partners' GDPs. From this, one can understand that bilateral trade is directly related to countries' size and inversely related to the distance between them plus any other factor (s) either aiding or resisting trade between the countries such as natural barriers (a land border), cultural barriers (common language) and various measures of manmade trade costs (free trade agreements). Therefore, a set of dummies incorporating some characteristics common to specific flows can be added to the model (Anderson 1979; Bergstrand 1989). Hence, like many empirical applications of the model and based on the main determinants of trade flows, the following log linear gravity equation is used:

$$\begin{aligned} \mathrm{LnT_{ij}} = {} & \beta_0 + \beta_1 \ln \mathrm{GDP_i} + \beta_2 \ln \mathrm{GDP_j} + \beta_3 \mathrm{EX_i} + \beta_4 \mathrm{EX_j} + \beta_5 \mathrm{Bord_{ij}} \\ & + \beta_6 \mathrm{Lang_{ij}} + \beta_7 \ln \mathrm{Dist_{ij}} + \beta_8 \mathrm{CFA_{ij}} + \beta_9 \mathrm{EAC_{ij}} + \beta_{10} \mathrm{FTA_{ij}} + \varepsilon_{ij} \end{aligned}$$

where T_{ij} is the volume of bilateral trade between the reporting country i and trading partner country j, β_0 is the constant of the model, GDP_i, GDP_j and EX_i, EX_j are the production levels and exchange rates of country i and j respectively, $Dist_{ij}$ measures the geographical distance between capitals of nations i and j, $Bord_{ij}$ and Langij, are dummy variables denoting that the trade between i and j is due to the fact that they have a common border and language, they take a value of 1 when both countries in the pair share borders and a common official language. Three specific dummies were also added—the CFA dummy which takes 1 when both countries in the pair use CFA, a common currency used in some western and central African countries, which helps capture the common currency trade effects; the EAC dummy variable to see whether the common membership in EAC may increase trade between partner states is also included; and the FTA dummy which takes a value of 1 when both countries are members of the same FTA, ε_{ij} is the error term. It is known that the bigger the countries in terms of their economic structures, the larger are the volumes of their bilateral trade. Therefore, β_1 and β_2 are expected to be positive. Currency depreciation (increase in exchange rate) in a country is expected to increase the exports of that country because the latter will be cheaper. Besides this, the appreciation of the currency (decrease in exchange rate) of the trading partner may reduce the exporter country's imports as imports will be expensive for the importer country. Thus, in view of this argument it is expected that β_3 and β_4 may take positive or negative values respectively. A common border and language means low cost and easier market access and thereby β_5 and β_6 are expected to have a positive sign. A larger distance between the countries is expected to have a negative impact on trade flows, thus β_7 is expected to have a negative sign, β_8 is also expected to be positive, showing a positive impact of using the same currency on bilateral trade. β_9 and β_{10} indicate an increase in the propensity to trade given that both nations belong to the same regional agreement (EAC and any other FTA). Hence, both coefficients are expected to have a positive sign (Bergstrand 1985; Baldwin and Taglioni 2006).

3.1 Analysis of Degree of Openness

Several methods, including the value added based openness indices as opposed to traditional indices can be applied to analyse the degree of openness. The latter is a simple and easy method that measures intra-regional trade by relating the value of goods and services sold and bought by the country to and from member states of an integration area to the value of all goods and services produced by domestic factors of production for domestic and foreign expenditure (that is GDP) for the period of 1 year and expresses it as a percentage. Thus, a value of zero per cent indicates a closed economy and the higher the value, the more the country is open to other member countries of the integration area (Kotcherlakota and Sack-Rittenhouse 2000).

The traditional method is most used when value added trade flows are lacking. Thus, given the available data and also that collected (which are given in gross terms) from 'IMF: Direction of Trade Statistics' (DOTS), the following tradition method equation has been used for analysis:

$$IO = \frac{(\text{sum of Exports and Imports})}{\text{Total GDP}} \times 100 \text{ with IO denoting the index of openness}$$

The empirical realizations of the degree of openness in 12 selected countries in the Euro Zone (which are likely to have experienced further trade integration given their early entry to EMU), are calculated and presented, and the same methodology is used for EAC countries as well. All the data are annual for both MUs.

3.2 Analysis of the GDP Business Cycles

The well-known HP filter[2] was used to estimate the cyclical components of each of the GDPs in the studied countries; the data are annual and cover 12 Euro Zone countries and five EAC countries from 1997 through 2013. After calculating the ratio of individual countries' GDPs to world GDP, the ratio was transformed in two ways: the natural logarithms of the GDPs' ratio were de-trended and the cyclical components of GDP in which we are interested were extracted.

4 Descriptive Statistics

4.1 Degree of Openess

4.1.1 Euro Zone Countries

The results from a study of the data helped us to represent the degree of openness for cross-country comparisons. Figure 2 illustrates the case of EMU countries.

We observe that during 1997–2013, Euro Zone countries experienced intra-union trade which was very high for Germany, high for some countries like France, Belgium, Netherlands and Italy, but countries such as Portugal, Austria, Ireland, Luxembourg and Greece experienced a very slow trade openness. One can say that a part of these differences is attributable to the different levels of economic activities in these countries such that a higher level of economic activity may lead to an increased volume of trade.

Overall, what we observe is a slight change in the level of intra-trade between these countries which kept the same features over time. This is the point that has been

[2] The Hodrick-Prescott (HP) filter is a specialized filter for trend and business cycle estimations.

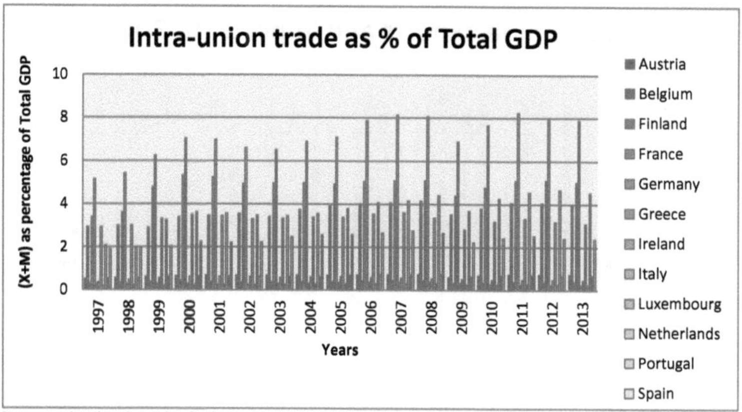

Source: Author's calculations based on IMFdata, DOTS.

Fig. 2 Degree of openness for selected EMU countries 1997–2013

stressed by the OCA theory—countries with less openness experience the cost of monetary union while those with a higher degree of openness enjoy the benefits of the union. This is exactly what happened in the Euro Zone where countries like Ireland, Portugal and Greece have been struggling to enjoy the benefits of the Euro Zone. Conversely, when we take the case of Germany, France and Belgium one can say that they have been enjoying their competitiveness in terms of benefits of being in EMU. This conclusion is again consistent with Baldwin (2006). In his study on the trade effects of the euro, as there was not enough data yet to answer the question 'how much did the euro boost trade?' he concluded that one needs to tackle the question 'if the euro boosted trade by sharpening competition, then in which dataset will we find the footprints?' For him, in any case, increased competition could go along with welfare gains even without observing increased trade flows (Baldwin 2006).

4.1.2 EAC Countries

Using the same methodology, the degree of openness among EAC countries was computed. Figure 3 shows that there was a wide difference in the degree of openness in EAC countries for more than a decade. Kenya had the highest degree of openness as compared to the other countries, Uganda and Tanzania also had a higher degree of integration and this is justified by the historical trade integration before Rwanda and Burundi joined EAC in 2007. Rwanda and Burundi had a small degree of openness with very small openness observed in Burundi's trade.

According to the optimistic view, and also on the basis of many of the conclusions from different researchers mentioned earlier, openness decreases asymmetric shocks. Hence, the more open a country the smaller will be the cost of a MU and the less open a country the greater will be the cost of a MU. Thus, a cost benefit analysis

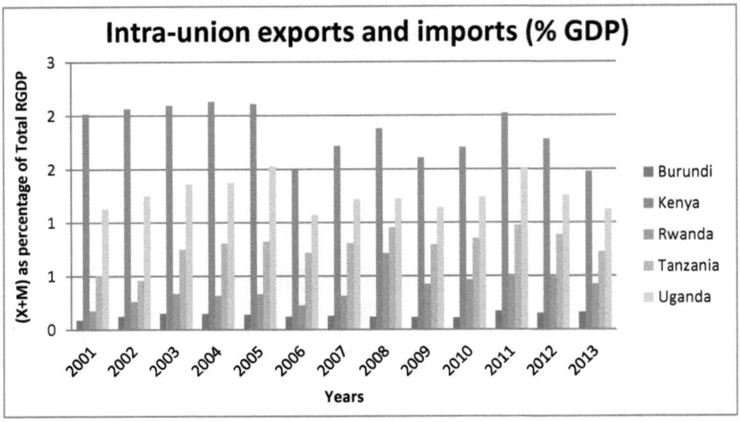

Source: Author's calculations based on IMF data, DOTS.

Fig. 3 Degree of openness for EAC countries 2001–2013

is likely to show net benefits of being in a MU for Kenya, Uganda and Tanzania; Rwanda and Burundi are not likely to benefit from a MU and therefore it is not clear whether they do belong to OCA ex-ante. However, given the OCA endogeneity hypothesis that has been justified by many of researchers, if these countries form a monetary union they are likely to deepen their intra-trade and therefore become an OCA ex-post, but throughout the overall process towards the EAC MU they should be warned by the current European experience. Hence, further studies should be conducted on this particular issue which can come up with clear conclusions on the endogeneity hypothesis with regard to EAC MU.

4.2 *Facts About Business Cycles*

4.2.1 Euro Zone Countries

The other variable for analysis was GDP business cycles for two groups of countries. For selected Euro Zone countries using data for the 1997–2013 period, a simple analysis of GDP_i/world GDP ratio[3] business cycles between countries revealed a similar co-movement, that is, little evidence of synchronization between GDP in all countries. This is consistent with the views of many authors, notably Camacho et al. (2006) who have developed indicators of the distance between national business cycles and have found that bilateral distances corresponding to the euro area countries tend to cluster together. This suggests that the business cycles of euro area countries have in fact much more in common with one another

[3] The ratio is calculated by taking each country i GDP divided by the world GDP.

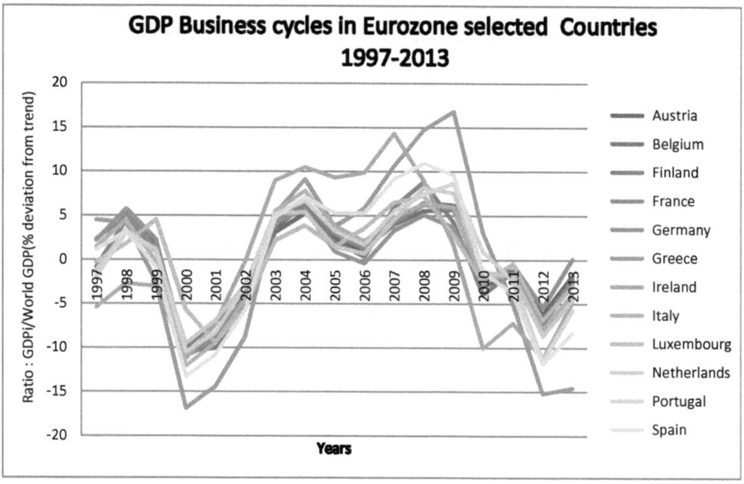

Source: Author's calculations based on WDI.

Fig. 4 GDP Business cycles in Euro Zone selected countries 1997–2013

than with other countries. Giannone and Reichlin (2006) also show that common shocks account for a bulk of the output fluctuations in Euro Zone countries. Countries' specific shocks in turn have small but persistent effects. Figure 4 shows the common features of GDP business cycles among 12 countries.

If you consider only Fig. 4, you may be tempted to conclude that there is strong evidence of synchronization of business cycles in EMU countries which is associated with the sharing of a common currency because as can be observed the highest co-movement is for the post-Euro Zone period, with an exception of course of after the global crisis, when in most countries the economic activities were different as were their economic performances. Nevertheless, given the differences in openness, it is not correct to conclude that the synchronization of business cycles between these countries emerged as a result of intra-regional trade as has been suggested by previous studies. This ambiguity needs direct empirical work for this particular issue, but as mentioned earlier, the relationship between openness and business cycles was raised by many researches and some of them also had to correct themselves. Economists like Bayoumi and Eichengreen (1993), carefully analysed business cycles and shocks affecting different potential EMU members to quantify the potential importance of a national monetary policy. But such an analysis implicitly takes business cycles' correlations to be unaffected by economic integration (Frankel and Rose 1998)

One notices that when considering the degree of openness, there is no clear conclusion with regard to the effects of trade on the symmetry of business cycles in the Euro Zone. However, we in fact conclude that similar movements of GDP business cycles in the Euro Zone may be a result of other factors that have become common to all countries as they have integrated more.

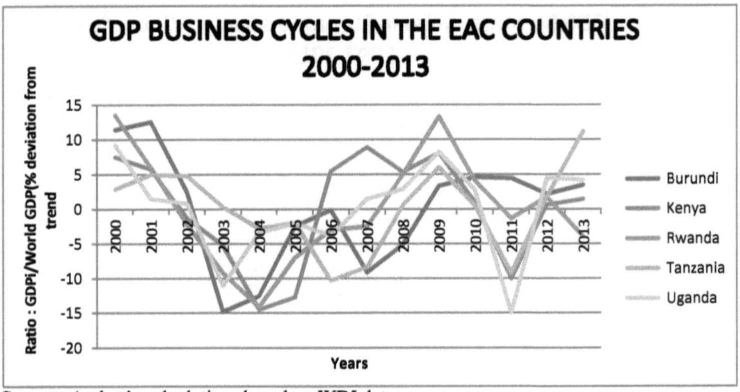

Source: Author's calculations based on WDI data.

Fig. 5 Business cycles co-movement in EAC (2000–2003)

4.2.2 EAC Countries

Using the HP filter, a co-movement analysis of GDP_i/world GDP ratio for EAC countries shows at first glance that movements of the economies of the EAC had similar features during the last decade, that is, they fluctuated in the same direction. However, there were times when some countries experienced different economic conditions (recession or booming), while others experienced similar patterns. This was the case in Tanzania and Burundi which experienced recession and a boom in 2006 and 2011 respectively. One may also notice that some amplitude of GDP business cycles is big (more volatile) while some is small (less volatile). Tackling the openness and volatility issue empirically, Giovanni and Levchenko (2006) found that countries that are more open to trade tend to be more volatile. They argued that this is the outcome of counteracting forces. Their findings are consistent with our analysis, which from trade openness results given earlier shows that countries like Kenya and Uganda are the most open to trade and they are also more volatile (see Fig. 5)

4.3 Conclusions from the Descriptive Analysis

The descriptive data was analysed to find out the possible positive relationship between trade integration and business cycle synchronization across selected Euro Zone countries and across EAC countries as previously shown by many studies. However, the analysis does not provide definitive and clear answers. On the one hand, both groups have shown differences in trade integration, and on the other hand they had a similar co-movement of business cycles which was tighter in Euro Zone countries than it was in EAC countries. This cannot be entirely taken as a result of trade intensity given that there might be other factors that have become common in all countries as they have integrated more. There is a need for a regression analysis so as to be able to show and confirm the links between trade integration and more symmetric business cycles. Future research on exploring the linkages between these variable is needed and expected.

5 Econometric Analysis

5.1 Data for Gravity Model

The dependent variable is the log of bilateral trade (sum of exports and imports) between pairs of countries (country i and j) in a given year. For simplicity of the model, cross-sectional trade data for 2000 and 2012 from DOTS were used; 53 African countries were included in DOTS (see Table 2 for the list of countries) but some of them do not show values of bilateral trade (perhaps because the level of trade is too small to be recorded or because countries did not report their respective trade flows), which leads to the problem of dealing with zero observations with the aspect of the gravity model in its linear form that uses a natural logarithm. The reason is that the standard procedure is to take the logarithm of the multiplication gravity equation so as to be able to estimate it in a linear form but one cannot take a log of zero. In addition, given that the use of the logarithm causes the model to lose many dependent variable observations (more than 700 observations), the OLS regression may lead to inconsistent parameter estimates, that is, biased results. To deal with this problem, different economists use different approaches, among which are a censored normal regression model and the well-known Tobit model (henceforth TM) which is commonly used (Frankel 1997). This is also the approach used in this paper which also uses two other approaches.

Data on explanatory variables were taken from different sources; GDP and exchange rate data from WDI; the World Bank dataset, dummy variables such as distance, language and borders were taken from the Centre d'Etudes Prospectives et d'Informations Internationales (CEPII) (Institute for Research on the International Economy in English); and information on membership to different FTAs and CFAs was taken from the Regional Trade Agreement database available on the WTO website.

5.2 What TM Does[4]

As many of the economists have pointed out (Wooldridge 2002; Sigelman and Zeng 1999; Frankel 1997; Greene 2012), in simple terms 'TM is a technique that estimates separate parameters to determine whether the observation is non-zero

[4] This part is mainly structured using Sigelman and Zeng (1999) and Greene (2012).

and then to determine what the coefficients are, conditional on the observation being non-zero' (Frankel 1997). This means that if some of the observations in the dependent variable denoted as y, equal zero, one should define a new random variable y which is transformed from the original one, y* (taken as latent variable). By:

$$y = 0 \text{ if } y^* \leq 0,$$
$$y = y^* \text{ if } y^* > 0.$$

To understand this concept better, it is like we are interested in population regression E(y*), if y* was observed for everyone in the population, OLS could be used. However, a data problem arises in that y* is not observed for some part of the population and causes censoring, that is, values on the dependent variable in a certain range of the dataset, are recorded to a single value, zero in our case. Hence, to handle that problem one can turn to TM with the general formulation which is usually given in terms of index function:

$$y_i^* = X_i \beta + \varepsilon_i$$
$$y_i = 0 \text{ if } y_i^* \leq 0,$$
$$y_i = y_i^* \text{ if } y_i^* > 0.$$

What we observe here is that the observed y_i is expressed in terms of y_i^*, a censored regression model that can be estimated using maximum likelihood estimation techniques. Thus, the log likelihood function for the TM used in this paper is:

$$ln\, L = \sum\nolimits_{i=1}^{N} \left\{ di\left(-ln\sigma + ln\phi\left(\frac{y_i - X_i \beta}{\sigma} \right) \right) + (1 - di) ln\left(1 - \Phi\left(\frac{X_i \beta}{\sigma} \right) \right) \right\}$$

This function is made up of two parts. The first is the classical regression for the uncensored observation while the second part corresponds to the probabilities that an observation is censured. Thus, TM represents three different expected values of which we might be interested in only one, that is the expected value of y_i conditional on y_i being greater than zero. That value is given with the function:

$$E[y_i y_i > 0] = X_i \beta + \sigma \lambda(\alpha)$$

where $\lambda(\alpha) = \frac{\phi\left(\frac{X_i \beta}{\sigma}\right)}{\Phi\left(\frac{X_i \beta}{\sigma}\right)}$

Hence, the marginal effects of x on y that we are interested in are the derivative of the function given earlier with respect to X_i, (view in the function given later) indicating how a one unit change in an independent variable x effects the dependent variable. Stata/SE13.0 were used to estimate the marginal effects of each of the independent variables on bilateral trade.

$$\frac{\partial E[y_i y_i > 0]}{\partial x_k} = \beta_k \left\{ 1 - \lambda(\alpha) \left[\frac{X_i \beta}{\sigma} + \lambda(\alpha) \right] \right\}$$

In addition, to deal with zero value entries, some researchers have often followed another strategy of simply omitting zero pairs from the dataset, and computing OLS estimates (see, for instance Frankel 1997). According to them, though, one must be concerned that the exclusion of zero values might bias the results. One could argue that a sample selection problem arises inevitably from leaving very small countries out of the dataset. They point out that even the few studies which include a broader set of countries leave out the smallest countries and they give an example of excluding countries like Andorra and Tuvalu (Frankel 1997). For the purpose of this paper, for example, no one will include Lesotho, Swaziland, Eritrea Namibia and Somalia, which are totally missing bilateral trade data in DOTS even though they are not the only ones. The same approach was used and the associated results are reported as OLS results. Another most commonly used strategy is expressing the dependent variable as log of (1 + Tradeij). Most of the authors who have used this approach argue that the logic is that on the one side, when the Tradeij is large the dependent variable is approximately equal to usual one lnTradeij, so the coefficients can be interpreted as elasticity in the usual way. On the other side when Tradeij is small the dependent variable is approximately equal to Tradeij itself which can be interpreted appropriately. This approach is also used and the associated results are reported as OLS 2 results. It is worth noting that all these ways that are used to deal with zero observations were just a simple robustness check.

5.3 Estimation Results

Table 1 gives the results from the three different approaches mentioned earlier (TM, regressing the gravity model using OLS by omitting zero pairs and by expressing the dependent variable as log of (1 + Tradeij)). Estimation results for both the periods (2000 and 2012) are presented separately.

Despite the drawbacks of each of the alternatives, their respective estimation results are somehow similar, that is, they generally have the same expected signs and levels of significance. These findings offer an assurance that the main determinants of trade in African countries are almost the ones that have been assumed previously. Table 1 also shows that almost all the explanatory variables have a highly significant effect (7 out of 10, including the constant of the model). On the one hand, the regressions reported in columns 4 and 5 show that the increase in exporter and importer country GDP promotes bilateral trade. For the Tobit coefficients we say that a one dollar increase in the exporting country's GDP increased bilateral trade by 0.0007 and 0.0011 dollar in 2000 and 2012 respectively and for the importing country the marginal effects varied between 0.0008 and 0.0012. Distance between partners reduces trade, a shared border, a shared language and

Table 1 Estimation results; Tobit vs OLS (dependent variable: bilateral trade)

	OLS[a]		Tobit		OLS 2[b]	
Variables	2000	2012	2000	2012	2000	2012
GDPi	0.8240***	0.7762***	0.0007***	0.0011***	1.6057***	1.4982***
	(16.63)	(16.80)	(12.68)	(13.88)	(18.83)	(18.01)
GDPj	0.8420***	0.8257***	0.0008***	0.0012***	1.5342***	1.4351***
	(17.55)	(18.35)	(13.66)	(14.46)	(18.47)	(17.55)
Exchange rate i	−0.0978***	−0.0300	−2560.1691*	−4394.3142	0.1077*	−0.0346
	(−3.00)	(−0.91)	(−1.74)	(−1.27)	(1.90)	(−0.56)
Exchange rate j	−0.0245	−0.0007	−895.7576	−4333.9323*	0.0991*	−0.0612
	(−0.78)	(−0.02)	(−1.48)	(−1.76)	(1.84)	(−1.04)
Border dummy	0.9094***	1.4554***	3.6565e+07***	1.9671e+08***	1.3867***	1.8188***
	(3.56)	(5.37)	(6.35)	(6.40)	(2.73)	(3.32)
Language dummy	0.7079***	0.5811***	1.7776e+07***	8.2681e+07***	1.6570***	1.9003***
	(5.18)	(4.13)	(5.64)	(5.04)	(6.93)	(7.38)
Distance ij	−1.5820***	−1.3119***	−6403.5706***	−14,850.3192***	−2.4548***	−1.8699***
	(−13.41)	(−10.83)	(−6.61)	(−3.02)	(−11.16)	(−7.92)
CFA dummy	0.6775***	0.8416***	−5.3192e+06	−7.5889e+06	1.0973**	1.2872**
	(2.62)	(3.23)	(−0.91)	(−0.24)	(2.26)	(2.56)
EAC dummy	1.0752	1.1510*	1.1824e+08***	9.7863e+07	0.6055	3.0229**
	(1.02)	(1.82)	(4.35)	(1.21)	(0.27)	(2.20)
FTA dummy	0.5850***	0.5746***	1.4402e+07***	9.0735e+07***	0.7380**	1.0184***
	(3.50)	(3.38)	(3.75)	(4.56)	(2.44)	(3.10)
Constant	−11.9928***	−13.4720***	−2.0547e+07***	−1.6665e+08***	−43.3039***	−44.0816***
	(−6.47)	(−7.17)	(−4.04)	(−6.43)	(−13.02)	(−12.48)
Observations	1476	1584	2218	2244	2218	2244
R^2/Pseudo-R^2	0.4471	0.4173	0.00977	0.00820	0.3469	0.3382

Note: t-statistics in parentheses: ***Significant at 1 % level, **Significant at 5 % level, *Significant at 10 % level

[a]Implies OLS estimates by dropping zero entries

[b]Implies OLS 2 estimates expressing the dependent variable as log of (1 + tradeij)

FTA membership increases trade as expected. For example, the coefficient on the distance variable says that the fact that if a partner country was 1000 km (unit measure of the distance variable) away from another country this reduced the bilateral trade by US$6403 and 14,850 dollars in 2000 and 2012 respectively.

Marginal effects of both FTA and EAC dummies reveal that countries which generally share memberships in the FTA and specifically those in EAC trade more between them than other country pairs. However, EAC estimates are not significant for 2012 compared to 2000 and one possible explanation for this is that in 2000 only three countries were members of EAC and two others joined in 2007 so perhaps it was too early for the trade effect to be shown among those five countries. Given the process towards full trade integration, it is worth noting that for OLS results, the insignificance shows up in 2000. Nevertheless, the signs are the ones expected. If one has to look at the Pseudo R^2 from Tobit results, one may be tempted to say that the model does not fit the data, which is not the case because the Tobit function is the log likelihood, that is, the log of a probability, so it can be negative or positive for continuous distribution or $0 \leq \text{Pseudo-}R^2 \leq 1$ for discrete distribution. So the pseudo-R^2, has no real meaning for Tobit,[5] instead one should look at chi squared which we found significant.[6] On the other side, the results in Table 1 columns 2 and 3, estimated using OLS by skipping/omitting zero values, can be interpreted as estimated elasticities. For example, coefficients for GDP show that increases in exporter and importer countries' GDPs increased bilateral trade with elasticities that varied between 0.776 and 0.842.

To be able to interpret the dummy variables' coefficients, one must use exponents of estimates for dummies. For example, border dummy coefficients of 0.9094 and 1.4554 imply that sharing a border magnified trade by $2.5 = e^{0.9094}$ to $4 = e^{1.4554}$ times higher than trade between countries that did not share borders in 2000 and 2012 respectively. The same interpretation can be used for all dummy coefficients; estimates of a FTA dummy imply that a pair of countries joining any FTA, trade about two times as much with each other than they do with countries not in the FTA. Western and central African countries using CFA, trade more compared to countries that do not use that currency—the factor was between 2 and 2.5 in 2000 and 2012 respectively and between 3 and 4 if we consider OLS2 results. This finding is consistent with empirical facts of studies carried out in the area of bilateral trade that have confirmed the positive effect of common currency on trade. However, estimations using the Tobit model showed the opposite situation, that is, countries in the CFA zone traded less compared to the group of countries not in the CFA zone. Again the elasticities for exchange rate for importers seemed to not be significant yet they had the expected sign. From this it can be inferred that an increase in exchange rates does not affect bilateral trade much in Africa. The R^2 indicates that the model explains more than 30 and 40 % of the variations in bilateral trade across African countries from both OLS results.

---

[5] Pseudo R^2 for Tobit, William Sribney, Statcorp, June 1997.

[6] Prob > chi2 = 0.0000 for 2000 and 2012.

5.4 Conclusion from the Gravity Model

With the exception of TM on CFA marginal effects, where it is shown that countries using the same currency trade less than countries using different currencies (this is unexpected under the gravity model and inconsistent with many of the studies that have been carried out in this area but we cannot pay much attention to this because the results are not statistically significant), the estimation results given now point to a number of findings and implications, from which one can confirm that the main determinants of bilateral trade in African countries are: levels of their production (GDP), sharing the same border and a common language, the distance between countries and membership in any FTA. Indeed, all the traditional gravity model variables have the expected sign and are highly significant.

6 Conclusion

Taking into consideration the existing literature, this paper tried to give answers to the question regarding trade effects of a MU within African countries, particularly within EAC partner states. Though it would be premature to draw a definitive conclusion, the paper throws up some interesting findings that need to be considered. Matching exercises on degree of openness and business cycle analyses using descriptive statistics and econometric analyses using cross-sectional data for 2000 and 2012, one may suggest that given that the EAC countries have been trading for at least a decade, the same period in which similarities in GDP business cycles are identified and that evidently, common memberships in FTAs have a significant positive effect on trade, continued EAC trade liberalization (through the MU that partner states are intending to form), can be expected to more positively and significantly impact bilateral trade as evidenced by CFA countries (OLS results) and lead to more synchronized business cycles. Therefore, it seems reasonable to conclude that a common currency is more desirable for EAC. Nevertheless, good economic analyses, that is, detailed and more convincing studies, are required for the desirability of an EAC MU to avoid serious macroeconomic problems that have so far been characterizing EMU. We expect to deal with more of these aspects in future research.

Appendix

Table 2 List of African countries from the DOTS database

Algeria	Djibouti	Libya	Senegal
Angola	Egypt	Madagascar	Seychelles
Benin	Equatorial Guinea	Malawi	Sierra Leone
Botswana[a]	Eritrea[a]	Mali	Somalia
Burkina Faso	Ethiopia	Mauritania	South Africa
Burundi	Gabon	Mauritius	Sudan
Cabo Verde	Gambia, The	Morocco	Swaziland[a]
Cameroon	Ghana	Mozambique	Tanzania
Central African Republic	Guinea	Namibia[a]	Togo
Chad	Guinea-Bissau	Niger	Tunisia
Comoros	Kenya	Nigeria	Uganda
Democratic Republic of Congo	Lesotho[a]	Rwanda	Zambia
Republic of Congo	Liberia	Sao Tome and Principe	Zimbabwe

Note: Sudan is taken as one country
[a]Countries do not have any bilateral trade values

References

Anderson JE (1979) A theoretical foundation for the gravity equation. Am Econ Rev 69:106–116

Adam AM, Chaudhry IS (2014) The currency union effect on intra-regional trade in economic community of West African states. J Int Trade Law Pol 13:102–122

Baldwin R, Taglioni D (2006) Gravity for dummies and dummies for gravity equations. Discussion Paper no. 12516, NBER, Cambridge, MA

Baldwin R (2006) The Euro's trade effects. Working Paper Series no. 594, ECB, Frankfurt

Bayoumi T, Eichengreen B (1993) Shocking aspects of European monetary unification. Working Paper no. 3949, NBER, Cambridge, MA

Bayoumi T, Eichengreen B (1997) Ever closer to heaven? An optimum currency area index for European countries. Eur Econ Rev 41:761–770

Bergstrand JH (1985) The gravity equation in international trade: some microeconomics foundations and empirical evidence. Rev Econ Stat 67(August):474–481

Bergstrand JH (1989) The generalized gravity equation, monopolistic competition, and factor-proportions theory in international trade. Rev Econ Stat 71(February):143–153

Calderon C, Chong A, Stein E (2007) Trade intensity and business cycle synchronization: are developing countries any different? J Int Econ 71:2–21

Camacho M, Pérez-Quirós G, Sáiz L (2006) Are European business cycles close enough to be just one? J Econ Dyn Control 30(9–10):1687–1706

De Grauwe P (1992) The economics of monetary integration, 1st edn. Oxford University Press, Oxford

De Grauwe P (2011) European monetary union. In: Durlauf SN, Blume LE (eds) The new Palgrave dictionary of economics. Palgrave Macmillan, Basingstoke

De Grauwe P (2014) The economics of monetary union, 10th edn. Oxford University Press, Oxford

De Grauwe P, Mongelli FP (2005) Endogeneities of optimum currency areas what brings countries sharing a single currency closer together? Working Paper Series no. 468, ECB, Frankfurt

EAC (2013) The protocol on the establishment of the East African MU. http://www.eac.int/index.php?option=com_content&view=frontpage&Itemid=1. Accessed 15 Jan 2015

Frankel JA (1997) Regional trading blocs in the World Economic System. Institute for International Economics, Washington, DC

Frankel JA, Rose AK (1997) Is EMU more justifiable ex post than ex ante. Eur Econ Rev 41:753–760

Frankel JA, Rose AK (1998) The endogeneity of the optimum currency criteria. Econ J 108:1009–1025

Giannone D, Reichlin L (2006) Trends and cycles in the euro area: how much heterogeneity and should we worry about it? Working Paper Series no. 595, ECB, Frankfurt

Giovanni J, Levchenko A (2006) Openness volatility and the risk content of exports. Meeting Paper no. 86, IFM, Washington, DC

Greene WH (2012) Econometric analysis, 7th edn. Pearson Education Limited, New York, NY

Horvath R, Komarek L (2002) Optimum currency area theory: an approach for thinking about monetary integration. Research Paper no. 647, TWERPS, Warwick University, Coventry

Jafari Y, Ismail MA, Kouhestani MS (2011) The determinants of trade among D8 countries: evidence from the gravity model. J Econ Coop Dev 32(3):21–38

Kotcherlakota V, Sack-Rittenhouse M (2000) Index of openness: measurement and analysis. Soc Sci J 37(1):125–140

Kouparitsas M (2001) Is the United States an optimum currency area? An empirical analysis of regional business cycles. Working Paper no. 22, Federal Reserve Bank of Chicago, Chicago, IL

Micco A, Stein E, Ordonez G (2003) The currency union effect on trade: early evidence from EMU. Econ Pol 18(37):315–356

Mkenda B (2001) Is East Africa an optimum currency area? Working Paper in Economics no. 41, Economics Department, Göteborg University, Gothenburg, Sweden

Mundell R (1961) A theory of optimum currency areas. Am Econ Rev 54(4):657–665

Rose A (1999) One money, one market: estimating the effect of common currencies on trade. Discussing Paper no. 2329, CEPR, London

Rose A (2004) A meta-analysis of the effects of common currencies on international trade. Working Paper no. 10373, NBER, Cambridge, MA

Sigelman L, Zeng L (1999) Analyzing censored and sample selected data with Tobit and Heckit models. Polit anal 8:167–187

Wooldridge J (2002) Econometric analysis of cross section and panel data. MIT Press, Cambridge, MA

Part III
Public Finances, FDI, Research, Innovation and Knowledge

Research, Innovation and Indigenous Knowledge in Sub-Saharan Africa: In Search of a Nexus

Chika Ezeanya

Abstract This paper attempts to establish a relationship between the low levels of innovation experienced across Africa south of the Sahara with the absence of indigenous knowledge in the education curriculum and in the continent's research and development agenda. Indigenous knowledge is the most easily accessible knowledge for most Africans; it is also a variant of knowledge of which several Africans have in-depth information. However, indigenous knowledge has been left out of classrooms and other organized teaching, learning and research platforms in the continent mainly due to the colonial foundations of education and the contemporary realities of continued dependence on external actors for education funding. Innovation, on its part, often occurs when an individual is equipped with in-depth and easily accessible knowledge of a particular field and/or locale. This study explores the concept of innovation and examines the experiences of nations with high levels of innovation, and establishes that indigenous or home-grown knowledge is foundational for innovation to thrive. It reaches the conclusion that the recognition of indigenous knowledge in formal, informal and non-formal education and research in Africa is foundational for creating a generation of Africans who are innovators and inventors, and who are self-motivated to conduct research on issues affecting the continent.

Keywords Innovation • Indigenous knowledge • Research, African development • Education

1 Introduction

Innovation, invention and creativity are the major drivers of growth and advancement in nations across the globe. A country that invests in creating an enabling environment for its human capital to operate at its optimum usually receives yields by way of highly innovative products and services. At the foundation of innovation

C. Ezeanya (✉)
College of Business and Economics, University of Rwanda, Kigali 1514, Rwanda
e-mail: chika@chikaforafrica.com

A. Heshmati (ed.), *Economic Integration, Currency Union, and Sustainable and Inclusive Growth in East Africa*, Advances in African Economic, Social and Political Development, DOI 10.1007/978-3-319-30432-8_6

and invention is an intimate knowledge of the environment within which the end-product will be used. Indigenous knowledge forms the basic foundation of knowledge for much of Africa's population south of the Sahara. Conversely, rather than ideally forming the foundation for teaching and research across the region, this knowledge has been marginalized from formal learning and research platforms. This state of affairs is traceable to the colonial origins of formal education and research in Africa and its continued dependence on external forces for education funding (Brock-Utne 2000). He who pays the piper dictates the tune; rather than an emphasis on Africa's indigenous knowledge in curriculum and in research focus areas, western curriculum and western agenda usually form the basis. This paper seeks to explore the relationship between the low levels of innovation and inventions in Africa and the absence of indigenous knowledge in teaching, learning and research across the continent. It starts by exploring the fundamental tenets of innovation and moves on to attempt a definition and conceptual clarification of indigenous knowledge, before looking at a relationship between the two by drawing examples from regions outside Africa. The conclusion is anchored on the assertion that an emphasis on indigenous knowledge is crucial in Africa's drive towards overall advancement and economic growth.

At its core, this paper is founded on G.R. Woodman and B. Morse's 1987 observation that the difficulty of designing viable development strategies in Africa derives from the fact that the region's modern development thinking is not the direct descendant from, or an adaptation of, the principles of the indigenous communities over which the new nation-states have imposed their rule. Research for development in Africa must not only be relevant to the needs of the people concerned and be appropriate to the social and material environments in which it is pursued (Hanushek and Ludger 2007), it must also be adaptive and cumulative, that is, it should respond to the exigencies of situations and be meaningful to members of society, taking into account their aspirations and concepts of development. In his 1999 *Development as Freedom*, Amartya Sen, opines that real development occurs when people are free to define their development based on societal dynamics. This view is also presented in many other variants by numerous scholars writing on education and societal advancement (see Brock-Utne 2000; Dewey 1959; Freire 1972, 1982). This paper explores the question as to whether incorporation of Africa's indigenous knowledge in the continent's research and development process will offer more viable approaches to its innovation and creativity across the region.

2 Innovation

Innovation generally entails the idea of doing new things. It is the whole process of renewing, changing, transforming or creating more efficient and effective means, products, processes or ways of doing things. There is widespread convergence around the fact that innovation is a major source of organizational or national wealth (Drucker 1992; West 2000). It has been said that innovation rules the world; nations that are constantly innovative have been shown to grow at a much higher rate as compared to nations that are rich in mineral, human or any other resources (OECD 2000). A good example is the United States, where it was assumed for over 200 years that economic growth came about as a result of inputs of capital and labor in the production process which resulted in greater output. However, Robert Solow, who later won a Nobel Prize in Economics for his work, was one of the few economists who discovered that only 15 % of the economic growth in the United States between 1870 and 1950 occurred as a result of increased input of labor and capital (Rosenberg 2004). That is, between 1870 and 1950, increased input of capital and labor 'could only account for about 15 % of the actual growth in the output of the economy. In a statistical sense, then, there was an unexplained residual of no less than 85 %' (Rosenberg 2004: 2). It was the unexplained residual of 85 % that 'persuaded most economists to conclude that technological innovation must have been a major force in the growth of output in highly industrialized economies' (Rosenberg 2004: 2).

Today, innovation in science and technology remains a major force in determining the rate of economic growth recorded by nations. In a study conducted by the United Kingdom's innovation foundation Nesta it was established that between 2000–2008, 63 % of the growth rate recorded in the country could be attributed to innovation, while only about 37 % could be linked to more inputs of capital and labor (Nesta Foundation 2013). According to the Foundation, research is the ability to convert ideas into needed novel products, services and processes and is the fountain of prosperity for developed countries (Nesta Foundation 2012).

Technological innovation is at the bedrock of the quest for improved economic growth in most nations across the globe. Innovation in several developed economies is a result of intentional, consistent and sustained investments in industrial and technological research by governments and the private sector (Grossman 1993). Technology implies the application of scientific knowledge, and often entails invention, innovation or the creation of a new product or method (Gordon and Waage 2010).

If investments in appropriate technologies are key to innovation, it is important to understand the concept of appropriate technology. For technology to be considered appropriate it must be founded on certain fundamental principles, which include:

- Accessibility and affordability
- Ease of utilization and maintenance
- Meeting the real needs of end users
- Effectiveness

Innovations in the field of technology, therefore, should keep these attributes in mind. The implication is that there is a need for deep knowledge of the environment where the product that is being developed is to be utilized. Researchers, inventors and innovators who have intimate understanding of their environments are often the ones who succeed in developing needful technology or other products, tangible and intangible, which impact the environment in deep and meaningful ways, oftentimes bringing about transformation and noticeable progress.

In a groundbreaking theory, Basu and Weil (1998) proposed that localized innovation is a strong and driving force in economic growth. According to their theory, 'new knowledge, although relevant for increased technological production can only be applicable or appropriate when used in countries whose previous technological productions are similar to the innovator's technology'. The implication is that when a product is developed in a particular environment, the innovation needed to improve on that product or develop offshoots from that product is more likely to be generated from the same environment where the original product was created. In essence, the idea of transferring technology is not sustainable since it is highly unlikely that imported technology will easily take root in a foreign environment and form a basis for more innovation in its new territory. It is in this regard that appropriate technology needs to be situated in the pre-existing technological knowledge or environmental reality of the innovator. This is where indigenous knowledge comes to the fore.

3 Indigenous Knowledge

Indigenous knowledge is local knowledge that has been distinctly generated from a particular society over a period of time. As distinct from the globally acknowledged knowledge that is often the product of research institutions, institutions of higher learning and privately owned firms, indigenous knowledge is 'the unique, traditional, local knowledge existing within and developed around specific conditions of women and men indigenous to a particular geographic area' (Grenier 1998: 1). Indigenous knowledge is specific to a particular region, while modern knowledge is usually seen as being universal in nature. Stilltoe et al. (2002: 9) defines indigenous knowledge as:

> Culturally informed understanding inculcated into individuals from birth onwards, structuring how they interface with their environments. It is also informed continually by outside intelligence. Its distribution is fragmentary. Although widely shared locally on the whole than specialized knowledge, no one person, authority or social group knows it all... It exists nowhere in totality, there is no grand repository.

Indigenous knowledge meets all the definitions of what appropriate technology entails, which as earlier spelt out are: accessibility and affordability; ease of utilization and maintenance; meeting the real needs of end users; and effectiveness. Indigenous knowledge is accessible to members of the community at little or no cost. It is very easy to use and maintain. That is, often there is little or no need for the importation of spare parts or expertise for indigenous knowledge to thrive and flourish. Indigenous knowledge, being knowledge that evolves in response to environmental challenges, has been established as being effective in meeting the needs of end users. In recognition of the importance and potential of indigenous knowledge in bringing about advancements within communities, the World Bank (2013: 1) asserts that:

> Significant contributions to global knowledge have originated from indigenous people, for instance in medicine and veterinary medicine with their intimate understanding of their environments. Indigenous knowledge is developed and adapted continuously to gradually changing environments and passed down from generation to generation and closely interwoven with people's cultural values. Indigenous knowledge is also the social capital of the poor, their main asset to invest in the struggle for survival, to produce food, to provide for shelter or to achieve control of their own lives.

During a presentation made in 2005 at the World Bank organized international workshop on indigenous knowledge in Benoni, South Africa, Marthinus Horak noted that the Indigenous knowledge system (IKS) although clearly different, is equal to Western knowledge systems. 'IKS may have ancient origins, but is relevant in day-to-day lives of people and continues to evolve, and is highly validated in context of community/local use' (Horak 2005: 3). It is important to emphasize Horak's point that indigenous knowledge is evolutionary and given to improvements. This is because indigenous knowledge is often viewed as traditional in nature, that is, static, immovable and resistant to modifications, expansions and transformations in tune with the times. To continue being relevant, indigenous knowledge must be able to adapt to changing conditions and environmental realities in the location where it is established.

Some scholars and researchers often use indigenous and traditional as one and the same. However, in the use of the word 'traditional', there is a tendency to restrict the meaning to a situation lacking in progress and something which is stagnant and even resistant to change. For Zartman (2000: 7), tradition should be viewed as a practice which has existed 'for an extended period and have evolved within societies rather than being the product of' exportation. Ake's (1990) use of the word traditional, however, bears more relevance to this paper. Ake defines traditional in its application to Africa as any form of knowledge and/or practice that is an authentic expression or outcome of Africa's history and intellectual evolution and experience. The implication here is that in Africa, indigenous knowledge is differentiated from western knowledge, which found its way into the continent as a result of contact with the west. Following the definition offered by Long (2004: 33), knowledge can be dichotomized in terms of 'modern science' versus 'people's science' or 'external' knowledge versus 'local' knowledge.

Historically, indigenous knowledge has been offered a lower place in the appreciation of knowledge globally. Terms such as 'primitive,' 'backward,' 'savage,' 'rural,' and 'unscientific,' to mention a few have been liberally applied in descriptions of indigenous knowledge for centuries. Often, a lack of 'universality', is ascribed to indigenous knowledge while only western science is arrogated a form of global acceptance and applicability (Kiggundu 2007: 49). For example, Brush and Stabinsky (1996), describes indigenous knowledge as being culture-specific, whereas western scientific knowledge is 'decultured.' However, as has been noted by several scholars, especially those of the critical theory and the post-modernist school of thought, 'knowledge is not and cannot be neutral either morally or politically or ideologically, as all knowledge reflects the interests of the observer' (Lemke 1994: 15).

4 Indigenous Knowledge and Innovation

Innovation is often a product of in-depth knowledge and expertise in a particular field. People who are well versed in indigenous knowledge and who also understand their environment intimately are more disposed to being inventors and being innovative if the right kind of support and access to information are made available to them. However, technology transfer or extension of technologies from developed, industrialized countries still thrives in the form of 'new crop varieties, medicine, fertilizers, computer hardware and applications and agricultural machines'. Although technology transfer has assisted Africa in some ways, it has also consistently proven not to be what can launch the continent to the level of advancement that it needs. Indigenous people often lack appropriate knowledge regarding the maintenance of imported technology and techniques. Africans, therefore, remain dependent on importers to upgrade the latest technology, making importation of technology unable to make any meaningful contribution to the GDP of Sub-Saharan Africa. On the other hand, indigenous technologies or knowledge have been developed and utilized over an extended period in Africa. They have been tried and tested within local communities and have proven to meet the immediate needs of the people. Examples include natural medicine, agricultural techniques and governance mechanisms. As a matter of fact, indigenous knowledge is known to have laid the foundations for quite a bit of what is considered modern knowledge or technology. In the United Kingdom, the efforts of Evan Thomas, a traditional bonesetter of enormous talent and training based in Liverpool whose satisfied clients included Prime Minister William Gladstone, transformed the field of traditional bonesetting to modern orthopedic medicine (Green 1999).

In China, starting of university level education in orthopedic medicine copied the curriculum available in western universities to begin with but it was not long before Chinese trained orthopedic medical practitioners opened up to indigenous knowledge of Chinese traditional bonesetters. These westernized, trained medical doctors have been able to redesign and severally re-modify following assumptions inherited from the west through their interaction with the knowledge of Chinese traditional bonesetters (Shang and Dong 1987). For example, the western method of dual plating was displaced by the bamboo split method used by China's traditional bone setters as this traditional bonesetter's treatment of displaced bone fractures of both the forearms was considered to be more superior (Fang et al. 1996). In Turkey, a country renowned for the effectiveness of the healing methods of traditional bonesetters who are found in abundance in the country, research in orthopedic medicine always factors in their knowledge. Several studies have been conducted and treatments certified by the nation's board as a result of collaborations with western trained orthopedic surgeons. In essence, the knowledge of Turkish traditional bonesetters is very much valued in research in modern orthopedic practice (Atici and Atici 2004: 50).

In the case of much of Africa south of the Sahara, it should be noted that indigenous knowledge about health and healing was solely relied on prior to the advent of the missionaries and the start of colonialism. The missionaries regrettably dismissed much of what is indigenously African as fetish, and demanded that converts rely instead on medication imported from Europe. Africans, convinced that it was entirely founded on sorcery began to disregard indigenous medical practices and rather patronized European medicine (Ahyi 1997). However, several Africans still believe in the efficacy of herbal remedies and traditional healing practices such as bonesetting. But notwithstanding its high patronage, traditional medicine receives little attention in Africa's educational system and research agenda. Institutional curricula, often founded entirely on western medical systems are closed and exclude the research and development of indigenous knowledge practices and systems.

5 Africa's Indigenous Knowledge in Practice

Indigenous knowledge cuts across fields and sectors and this section briefly delves into Africa's indigenous knowledge and its uses in select fields.

6 Indigenous Agriculture

During pre-colonial and colonial times, British missionaries and later colonial masters, coming with a supposedly superior knowledge of farming techniques sought to impress local African farmers to change their centuries-old method of farming. From the time of colonial incursions into Africa, indigenous knowledge of agriculture became the preserve of ignorant rural peasants. Thomas Odhiambo (1990) asserts that the education system in Africa has consistently ignored the knowledge, skills and survival strategies of local farmers, who have successfully managed their farmlands and remained productive for centuries with little or no external inputs.

Despite the glaring difference between Africa's tropical climate the subject of agriculture as taught in African schools and colleges relies almost entirely on European models and applications. Agricultural textbooks from which pupils are taught are in most cases imported from Europe; when they are published in Africa, they rely on the published works of European texts for ideas. There is little effort by African agricultural scientists to research the indigenous knowledge of rural farmers, and to use it in conjunction with borrowed knowledge from elsewhere in formulating science based, technological agricultural practices that would work effortlessly in Africa.

In teaching farmers across much of Sub-Saharan Africa's rural communities, emphasis has been placed on the importation of western knowledge of farming as a much more superior form of knowledge to be copied as a whole, without criticism or questioning. European agricultural technology is forced upon unlettered peasants, while their indigenous, and often sustainable and climate friendly farming and post-harvest techniques are dismissed as backward and untenable as far as improved agricultural practices are concerned. On the other hand, the western world is growing to the realization of the hazardous effects of certain modern farming practices such as the use of synthetic fertilizers, pesticides and herbicides. As a result, more and more westerners are seeking nature based or organic solutions for meeting the challenges of increased crop production, pests and weeds, in addition to post harvest processing and storage. In fact, researchers from the global North are increasingly diverting attention to rural areas in Africa in search of indigenous knowledge as a 'major untapped source for developing sustainable agriculture' (Odhiambo 1990: 3). Odhiambo (1990: 3) also states that, 'Indigenous knowledge can reveal missing ecological keys which may help scientists develop alternative agricultural technologies less dependent on non-renewable resources (e.g., fossil energy) and environmentally damaging inputs (e.g. chemical pesticides) than conventional technologies.'

There are numerous examples where indigenous knowledge in agriculture has contributed immensely to the development of the sector in Africa. Indigenous agricultural practices have been immensely beneficial in building food security across several communities around the globe. In schools and research institutes in Canada, for instance, the years of experience of the aboriginal people in managing their environment and determining appropriate planting seasons and soil knowledge is a huge resource for researchers and policymakers.

There are several instances where indigenous knowledge in agriculture has resulted in increased output in the area of food production across Sub-Saharan Africa. In Kenya, a colonial mandate of 1936 required every Kikuyu farmer to stop the practice of mixed cropping, which colonialists considered untidy. In its place, European style parallel-row mono-cropping was imposed. This adopted practice led to Kikuyu farmers losing a lot in terms of yield, and also being exposed to immense production risks and severe environmental degradation (Warren 1991: 3). However, about 50 years later Warren notes that at the global and national levels, agricultural research centers have been able to scientifically prove the effectiveness and efficiency of Kikuyu's indigenous mixed cropping systems. However, in the Kenyan education curriculum, the British imposed model is still being widely studied.

In Chad, local farmers had been using the indigenous zai rain-fed irrigation to successfully grow crops in the patched desert lands which they occupy. However, in a bid to modernize farming, the government of Chad borrowed heavily from the World Bank to finance modern irrigation agriculture in trying to achieve its food security goals. An initial attempt at commercial irrigation was unable to provide the necessary moisture to produce the needed tonnage of food. Farmers and communities became wary of submitting their lands for another project that might likely fail (the World Bank 1989). The farmers insisted that if given an opportunity, a modified and upgraded zai method would be able to benefit the nation a lot more than attempts at commercial irrigation. The World Bank conducted a financial and economic rate of return analysis on the 'cultivation of rice, wheat and sorghum, to determine the difference between the traditional rain-fed irrigation system and the more modern methods' (the World Bank 1989: 5). At the end of the analysis, it was revealed that in comparison the upgraded indigenous irrigation method yielded more economic profitability, especially with regard to cereal cultivation, particularly wheat and sorghum, than commercially irrigated cereals. The World Bank went on to advise that governments should carefully consider the options available to them locally, before taking a decision to sink funds in commercial irrigation systems, 'which are extremely costly in terms of both investment and operating costs' (the World Bank 1989: 28). The World Bank noted in making further recommendations, that, 'Governments and donors have tended to assume that farmers were interested in irrigated agriculture and failed to develop an understanding of how irrigation fits into the farmers' economic strategy' (the World Bank 2013: 5).

Regardless of the profundity of the study and its implications for transforming the face of farming in Africa and around the world, universities and research institutes in Chad are not known to be seriously involved in researching this unique indigenous irrigation method for up-scaling 'it to ensure increased agricultural productivity. On the contrary, the Chadian agriculture curriculum promotes government investment in commercial irrigation as a way to increase yield and combat food insecurity' (Ezeanya 2011: 128). Other examples include Neem bio-pesticides in Togo and Niger and ethno-veterinary medicine and fishing in the Niger River (Warren 1991: 14).

7 Indigenous Knowledge of the Environment

African communities have a deep and intimate understanding of their environment, but this knowledge is not factored in when conducting research, environmental impact assessments of infrastructural projects or in designing policies that affect the environment. Indigenous knowledge of the environment implies a thorough understanding of the lifecycle, development processes, location and other relevant details of the community terrain, including plant life, animal kingdom and natural phenomena (Appiah-Opoku 2005: 103):'Indigenous knowledge of the environment goes beyond its mastery to emphasize the fashioning of appropriate technologies for utilizing environmental resources in a sustainable manner.' Among the Khoisan people, a predominantly hunter-gatherer community found in parts of Southern Africa, indigenous knowledge of the habitat is extensively used with high accuracy to determine the movement of animals and any change in weather conditions. Indigenous knowledge held by African communities is not generally known to western scientists and western knowledge systems. However, African students and researchers are taught with western curricula and are guided by western agenda and yardsticks/background information (Knudston and Suzuki 1992).

Indigenous knowledge is also much needed in doing environmental impact assessments (EIAs) across African countries. The existing western based EIA models have been criticized for being 'technical, reactive and narrow in scope of application', often rendering them inflexible and unable to accurately assess the impact of infrastructural projects on the environment (Appiah-Opoku 2005: 18). Western EIA models rely heavily on assumable and predictable socioeconomic and political conditions which are not usually applicable to much of Africa. Indigenous knowledge based EIAs will be much more appropriate for understanding Africa's terrain as they have been beneficial in several projects in the territory of the aboriginal Canadians, such as Beaufort Sea Hydrocarbon Production and Transportation, the Oldman River Dam and the Norman Wells Oil Field Development and Pipeline Project (Berkes 1988). African governments and researchers can borrow a leaf from Canada's experiences in instituting indigenous knowledge based EIA procedures for Africa and insisting on them, especially with donor funded projects where western consultants are usually brought in to assess the impact of projects.

8 Indigenous Knowledge of Trade and Savings

Capital as defined by economists implies any commodity whose value is generated from its use in the production of other goods. Machines, funds and other assets such as tractors or trawlers need capital to ensure continued outputs. In its popularized variant, capital is referred to as 'funds or money needed to operate or to start a business' (Ayittey 2006: 344). Indigenous Africa relies on communal efforts to raise capital for large scale projects. This is done through revolving credit schemes known by different names in different parts of the continent: *susu* in Ghana, *esusu* in Yoruba, *tontinnes* or *chilembe* in Cameroon and *stokfel* in South Africa (Illiffe 1987: 136).

> Typically a group of, say, 10 people would contribute, say, $100 into a fund. When it reached a certain amount, say $1000 it was handed over to the members in turn. Such a scheme required a liberal dosage of trust among members to be operational and somehow the natives managed to make it work. In fact, for many businesses in the indigenous and informal sector, the loan club was their primary source of capital (Ayittey 2006: 344).

To borrow money, individuals pledged their farms or in case they could not do so, formed a partnership with someone else who had the capital needed to execute the project. Further, a unique system of trading in Africa is the institution of trust. In the area of commerce, it was possible for 'middlemen or agents to secure credit solely on the basis of trust' (Ayittey 2006: 345). The institution of trust operated in advancing goods to a trader by a producer or importer, the repayment of which was expected from the latter within an agreed timeframe and in a medium that was acceptable to the supplier:

> At Old Calabar in 1851, the British Council estimated that at least 70,000 pounds (sterling) of imported goods were in the possession of brokers and a further 13,000 pounds (sterling) had been advanced and already traded to suppliers. Another observer found that 'with the utmost confidence a fellow nearly naked will ask you for three, four, or even five thousand pounds (sterling) worth of goods on credit, and individuals are often trusted to that amount.

> I have trusted more than one man with goods, the returns of which were worth between two and three thousand pounds.' Trust formed the essential part of the agreements between Sierra Leone traders and King Docemo of Lagos in1854... Gambia, the scale of trust in the 1850s was about 200 to 2,000 pounds sterling per agent, and there were eight or ten agents for each French firm (Newbury 1971: 19).

In West Africa as well in the nineteenth and early twentieth centuries, palm-oil trade dominated all other forms of legitimate trade. This trade was run on the same unique and indigenous knowledge based system of trust that existed between European merchants and the African middlemen in the coastal areas who purchased palm-oil from the hinterlands, which were inaccessible and unknown to Europeans. African middlemen greatly benefited from this indigenous knowledge based system as related by Newbury (1971: 19):

> An African was compelled to sell all his oil to the European whose trust he held. The European never wanted his trust totally repaid by a reliable merchant because the African would then be free to sell to the European's rivals. Europeans tried every method, honest and dishonest, to keep Africans in debt to them. To break the monopoly hold on Africans, new firms would offer either higher prices or trust on easier terms. If the Africans supplied the new merchant with oil the old firms would forcibly seize it. The king would then declare a boycott of all trade until the dispute was settled. The king also declared a trade boycott when the European firms combined to fix prices. Nevertheless, despite its imperfections, the trust system did supply Africans with some credit to begin commercial operations.

9 Indigenous Pharmacology

Indigenous pharmacology is a term that refers to knowledge about healing properties of plants, roots, barks, animal products and other naturally occurring substances that is held by communities. Among African communities, knowledge about the healing properties of naturally occurring resources abound; diseases such as common cold, fever, sores, diabetes, malaria and fractured bones have been known to be remedied with knowledge passed on from generation to generation (Baronov 2008).

Western bio-medicine has undoubtedly contributed enormously to the treatment and management of several diseases in the past century and also in recent times, but with an exponential increase in diseases that defy diagnosis and remediation through western medicine, scientists are beginning to look at indigenous medicine. African countries hold great potential for the discovery of herbs, seeds, animal sources, trees and even clay with healing properties. Some of the more recently established and scientifically proven African indigenous remedies that have been incorporated into western medicine include South African *Hoodia Gordini* for the treatment of obesity (Konadu 2007); type II diabetes management drug from Kenya (McGown 2006) and Iboga (*Tabernantheiboga*) a plant that has been used as a stimulant for centuries in Central and West Africa which has now been established as a cure for addiction (McGown 2006), to mention a few.

10 Indigenous Mathematics

African mathematics is a much overlooked field of study, but it is one which holds enormous potential for innovation across disciplines. Mathematics is an everyday field and is utilized by most adults at work, at home and in leisure in varying degrees. Traditional African societies are no exception to this. Teaching of mathematics across Africa has continued in the European tradition established by missionaries and consolidated during the colonial era. Apart from the much studied ancient Egyptian mathematics, mathematics from several communities in Africa south of the Sahara has received minimum attention from academia.

Few African students know that some of the earliest mathematics objects in human history were discovered in Africa. The Lebombo bone, dated approximately 35,000 BC was discovered in the mountains of South Africa and Swaziland, while the Ishango bone dated 6000–9000 years was discovered on the border of Uganda and the Republic of Congo (Bangura 2012). In African art such as textiles, wood carvings and mural decorations can be found consistent and in-depth geometrical expressions. Africa's numbering system also displays a surprising similarity across several ethnicities and cultures providing a strong platform for further research (Bangura 2012).

Several indigenous African activities, games, end products and manufacturing processes are dense with lessons from mathematical sub-fields such as fractals, combinatorics, bifurcation, tiling or tesellation (Bangura 2012). In the sub-field of African fractals, for instance, American researcher Ron Eglash has done the most work on the use of fractal patterns in African architecture, art and religion (Eglash 2002). According to Eglash, in Africa he encountered 'some of the most complex fractal systems that exist in religious activities such as the sequence of symbols used in sand divination, a method fortune telling found in Senegal' (Bangura 2012: 6). In *African Mathematics*, Abdul Karim Bangura notes that although rarely studied, numerous African indigenous games involve combinatorics. African board games, for example, hold much promise for research as they are 'games of strategy, full of information, logic and intelligence [and therefore] it is imperative to ask questions of intelligence, logic and mathematical reasoning when investigating them' (Bangura 2012: 79).

11 Indigenous Knowledge and Research in Africa

There is an obvious disconnect between what is generally researched in Africa and what the real needs of a majority of the Africans are. G.R. Woodman and B. Morse (Woodman and Bradford 1987) have observed that it has proven difficult to design workable development strategies in Africa due to the fact that the region's contemporary approach to development is a complete deviation from the knowledge, principles and values of the indigenous communities over which the colonially contrived nation-states have imposed their rule. Education is the surest and quickest path to ensuring social continuity and bringing about transformation in any society. For

Dewey (1959: 3), human beings 'are born not only unaware of, but quite indifferent to, the aims and habits of the social group and have to be rendered cognizant of them and made to become actively interested; education and education alone, spans the gap.' Education ought to be based on the real-life experiences of learners and what their immediate environment and social realities entail. In this instance any 'rift between curriculum and society must be bridged' (Walker and Jonas 1986: 11).

The World Bank (2005: 1) admits that 'educational research has shown that teaching supported with prior knowledge increases students' ability to grasp materials taught to them (...) and they are more apt to retain information. In Sub-Saharan Africa, education and research has mostly taken the form of an outside-in approach whereby the agenda of what is to be researched is set by donors or development partners. This is also the case with teaching and learning curricula. Very few efforts that are transformative in approach and content have been made to modify the curricula of teaching and learning across the continent of Africa to make for independent, environmentally generated and sensitive teaching, learning and research.

Research that will lead to advancement in Africa will have to be founded on appropriate education. Classroom content must integrate 'particular curriculum content and design, instructional strategies and techniques, and forms of evaluation' (Trifonas 2003: 23). In Africa, research agenda, curriculum and 'given' conceptual frameworks should be continuously re-examined by researchers, teachers and students with the aim of eschewing all manifestations of neo-colonial underpinnings and emphasizing indigenous ideas and addressing Africa's peculiar realities and challenges (Ezeanya 2011).

In the search for knowledge within any particular community, people's history, culture and worldview ought to form the baseline for further studies and analysis (Sarpong 2002). Africa is rich in indigenous knowledge in all fields and sectors, which the advent of western methods of scientific inquiry has repudiated. The result of a disregard for what is authentically African in agriculture, science, mathematics, geography, arts, medicine, politics and economics, to mention a few is a detachment of research from the people's lived experiences. African researchers struggle for relevance and to have the masses appreciate their research outputs, but this has proven difficult over the years as a result of the disconnect that exists between research and reality. According to Mkabela (2005), it is important that the African reality be examined from an African perspective where the African, his experience and voice are located in the centrality of the research and are constantly scrutinized and reaffirmed. For the creation of a dynamic multicultural approach to research, therefore, it is very important for African researchers to reacquaint themselves with Africa's knowledge systems and research.

There is need for African researchers to merge the western acquired knowledge, skills, methodologies and tools of research with the African reality (Nsamenang 1995). In essence, western solutions and research strategies for discovering new knowledge are not made to measure for all. The west does not hold the key to research methodologies and approaches for understanding the rest of the world. In Africa, lack of emphasis on this truth has brought about distortions in efforts towards advancing the continent and its people. According to UNESCO (2000:

1), 'new insights reveal that development interventions have failed to induce people to participate because of the absence of instruments and mechanisms that enable them to use their own knowledge. Greater efforts therefore should be undertaken to strengthen the capacity of local people to develop their own knowledge base and to develop methodologies to promote activities at the interface of scientific disciplines and indigenous knowledge.' African researchers are reluctant to tackle the challenges facing the continent unless they are funded or enter into some sort of partnership with western institutions. These are hindrances to real research work in Africa and the production of authentic knowledge out of the continent.

12 Conclusion

Despite decades of graduating university and post-graduate students in Africa, a low level of innovation and invention across the region has been recorded. Persistence of several developmental challenges and the snail speed movement towards technological advancements across Africa point to foundational issues with the region's research and development agenda. Africa, for instance, records the lowest number of patent applications in the whole world. This begs the question: how are the existing curricula and research agendas contributing to the dearth of widespread innovations and inventions across the continent? Inventions, innovations and creativity spring from a place of familiarity and spontaneity. Indigenous knowledge is the knowledge that many Africans are acquainted with, but which is not incorporated in teaching curricula and research agenda across the continent. Research for an average African researcher is an enigma that can only be unwrapped with funding, assistance or publication platforms offered outside of the continent's realities.

Rich and potentially life transforming indigenous knowledge in certain fields explored in this paper such as agriculture, environment, pharmacology and mathematics if incorporated as part of the curriculum of teaching and as a foundational part of the research agenda in Africa, will most likely result in a remarkable increase in innovation and creativity across the region. To work effectively with indigenous knowledge, researchers will definitely have to be 'a little more humble, patient, determined, sensitive, flexible, creative, unconventional, open minded, critical, and cautious; a commitment to positive social change and to conducting enriching and enhancing research is needed' (Grenier 1998: 42). Empowering indigenous knowledge as a fundamental aspect of research in Africa will make for spontaneity which has been identified as the bedrock of creativity and innovation. Ease of access to research materials will be another major reason why innovation and creativity will thrive in Africa when indigenous knowledge is emphasized. Transformative home-grown and grassroots based creativity and innovation across all sectors including trade and economics, philosophy, science and technology, the creative arts, politics and social and cultural will be experienced when Africa's indigenous knowledge is accorded its place in research.

References

Ahyi G (1997) Traditional models of mental health and illness in Benin. In: Hountondji P (ed) Endogenous knowledge: research trials. CODESRIA, Dakar

Ake C (1990) Sustaining development on the indigenous. Long-term perspective study of Sub-Saharan Africa, Background Papers, Volume 3: Institutional and sociopolitical issues. The World Bank, Washington, DC

Appiah-Opoku S (2005) The need for environmental impact assessment: the case of Ghana. The Edwin Mellen Press, New York, NY

Atici E, Atici T (2004) The development of orthopaedics and traumatology in Turkey and some results. JISHIM 3:50–55

Ayittey G (2006) Indigenous African institutions. Transnational Publishers, New York, NY

Bangura A (2012) African mathematics: from bones to computers. University Press of America, Lanham

Baronov D (2008) The African transformation of western medicine and the dynamics of global cultural exchange. Temple University Press, Philadelphia, PA

Basu S, Weil D (1998) Appropriate technology and growth. Q J Econ 113(4):1025–1054

Berkes F (1988) Environmental philosophy of the Chisasibi Cree people of James Bay. In: Freeman M, Carbyn L (eds) Traditional knowledge and renewable management. Boreal Institute for Northern Studies, Edmonton, pp 201–250

Brock-Utne B (2000) Whose education for all: the recolonization of the African mind. Falmers Press, New York, NY

Brush S, Stabinsky D (1996) Valuing local knowledge: indigenous peoples and intellectual property rights. Island Press, Washington, DC

Dewey J (1959) Moral principles and education. Philosophical Library, New York, NY

Drucker P (1992) Managing for the future. Butterworth-Heinemann, Oxford

Eglash R (2002) African fractals: modern computing and indigenous design. Rutgers University Press, New Brunswick, NJ

Ezeanya C (2011) Education and indigenous knowledge in Africa: traditional bonesetting and orthopaedic medicine in West Africa. Howard University, Washington, DC

Fang H, Ku Y, Shang T (1996) The integration of modern and traditional Chinese medicine in the treatment of fractures; a simple method of treatment for fractures of the shafts of both forearm bones. Clin Orthop 323:4–11

Freire P (1972) Pedagogy of the oppressed. Penguin Books, Harmondsworth

Freire P (1982) The politics of education: culture, power, and liberation. Bergin & Garvey, South Hadley, MA

Gordon C, Waage J (2010) Science and innovation for development. United Kingdom Collaborative on Development Sciences (UKCDS), London

Green S (1999) Orthopaedic surgeons: inheritors of tradition. Clin Orthop 363:258–263

Grenier L (1998) Working with indigenous knowledge: a guide for researchers. IDRC, Ottawa

Grossman G (1993) Innovation and growth in the global economy. MIT Press, Boston, MA

Hanushek E, Ludger W (2007) The role of education quality in economic growth. World Bank Policy Research Working Paper, WPS4122, (February). The World Bank, Washington, DC

Horak M (2005) Adding value to indigenous knowledge through scientific innovation. Retrieved from World Bank: http://www.worldbank.org/afr/ik/GRA/horak.pdf

Illiffe J (1987) The African poor. Cambridge University Press, New York, NY

Kiggundu J (2007) Intellectual property law and the protection of indigenous knowledge. In: Mazonde I, Thomas P (eds) Indigenous knowledge systems and intellectual property in the twenty-first century: perspectives from Southern Africa. Codesria, Dakar, pp 26–47

Knudston P, Suzuki D (1992) The wisdom of the elders. StoddartLePena, Toronto, ON

Konadu K (2007) Indigenous medicine and knowledge in African society. Routledge, New York, NY

Lemke J (1994) What is postmodernism and why is it saying all these terrible things. J Accel Learn Teach. http://academic.brooklyn.cuny.edu/education/jlemke/papers/jsalt.htm Accessed 16 Nov 2007
Long N (2004) Contesting policy ideas from below. In: Boas M, McNeill D (eds) Global institutions and development: framing the world? Routledge, London
McGown J (2006) Out of Africa: mysteries of access and benefit sharing. Edmonds Institute Report, Washington, DC
Mkabela Q (2005) Using the Afrocentric method in researching indigenous African culture. Qual Rep 10(1):178–189
Nesta Foundation (2012) Plan I—the case for innovation led growth. Nesta, London
Nesta Foundation (2013) Innovation drives economic growth. Retrieved from A Nesta Foundation website: http://www.innovationexcellence.com/blog/2013/06/29/innovation-drives-economic-growth-news-from-nesta/#sthash.SYf2elRr.dpuf
Newbury C (1971) Prices and profitability in early nineteenth-century West African trade. In: Meillassoux C (ed) The development of indigenous trade and markets in West Africa. Oxford University Press, Oxford
Nsamenang A (1995) Factors influencing the development of psychology in sub Saharan Africa. Int J Psychol 30(6):729–739
Odhiambo T (1990) You cannot fix indigenous knowledge. ILEIA 6(1):3–5
OECD (2000) Mobilizing human resources for innovation. OECD, Paris
Rosenberg N (2004) Innovation and economic growth. OECD, Paris
Sarpong P (2002) Peoples differ: an approach to inculturation in evangelization. Sub-Saharan Publishers, Legon
Sen A (1999) Development as freedom. Oxford University Press, Oxford
Shang T, Dong Y (1987) Treatment of forearm bone fractures by an integration of traditional Chinese and Western medicine. Clin Orthop Relat Res Feb(215):56–64
Stilltoe P, Bicker A, Pottier J (2002) Participating in development: approaches to indigenous knowledge. Routledge, New York, NY
Stilltoe P, Bicker A, Pottier J (2005) Indigenous knowledge notes. Retrieved from the World Bank: http://www.worldbank.org/afr/ik/ikn87.htm
Stilltoe P, Bicker A, Pottier J (2013) What is indigenous knowledge. Retrieved from the World Bank: www.worldbank.org/afr/ik/basic.htm
The World Bank (1989) Chad irrigation subsector review. The World Bank, Agricultural Division, Sahel Department, Africa Region, Washington, DC
Trifonas P (2003) Toward a decolonizing pedagogy: social justice reconsidered. In: Trifonas P (ed) Pedagogies of difference: rethinking education for social change. Routledge, London
UNESCO (2000) Best practice on indigenous knowledge. Retrieved from United Nations Education Scientific and Cultural Organization: http://www.unesco.org/most/bpikpub.htm#role
Walker D, Jonas S (1986) Curriculum and aims. Columbia, New York, NY
Warren D (1991) Using indigenous knowledge in agricultural development. The World Bank, Washington, DC
West J (2000) The mystery of innovation: aligning the triangle of technology, institutions and organisation. Aust J Manag, 26, Special Issue, August: 21–43
Woodman GR, Bradford M (1987) How state courts create customary law in Ghana and Nigeria. In: Woodman G, Bradford M (eds) Indigenous law and the state. Fows Publications, Dordrecht, Holland
Zartman IW (2000) Introduction: African traditional conflict “medicine”. In: Zartman IW (ed) Traditional cures for more conflicts: African conflict ‘medicine’. Lynne Reinner, Boulder

The Shadow Economy and Corruption as Development Impediments

Almas Heshmati

Abstract Shadow economic activity has been on the rise leading to many problems for society, the state and also when it comes to trust in its organizations. The shadow economy has a significant share of the overall economy and captures all the activities beyond those measured by official activities. The rise of the shadow economy around the world is attributed to the stronger presence of government activity, increase in tax rates and taxpayer firms and households' desire to escape taxes and regulatory restrictions. This paper is a review of recent studies investigating various theoretical and empirical aspects of activities in the shadow economy and the measurement and development of the shadow economy across developed, developing and transition economies. It discusses a number of areas related to the shadow economy and their relationship with public performance and economic growth. The review leads to the identification of several indicators with negative or positive association with the size of the shadow economy and its causal effects.

Keywords Shadow economy • Tax morale • Institutional quality • Corruption • Economic freedom • Economic growth • Effective democracy • Natural resources • Political culture

1 Introduction

Activities in the shadow economy have been on the rise and are leading to violations of laws and regulations, lowering tax revenue collections, statistical discrepancies, inequalities, corruption, public budget deficits and public debt problems for the state and its organizations. The shadow economy captures all the

---

For a full but different version see Chap. 3 in Tausch et al. (2014).

A. Heshmati (✉)
Jönköping International Business School, Jönköping, Sweden

Department of Economics, Sogang University, Room K526, 35 Baekbeom-ro, (Sinsu-dong #1), Mapo-gu, Seoul 121-742, Korea
e-mail: heshmati@sogang.ac.kr

A. Heshmati (ed.), *Economic Integration, Currency Union, and Sustainable and Inclusive Growth in East Africa*, Advances in African Economic, Social and Political Development, DOI 10.1007/978-3-319-30432-8_7

activities that are beyond the measurement of official activities. In all countries, there is evidence that the shadow economy has a significant share of the overall economy. It is also labeled as hidden, black, underground, unobserved, unofficial, unrecorded and parallel economy. The rise of the shadow economy around the world is attributed to the stronger presence of government activity, increase in tax rates and the desire to escape taxes and regulatory restrictions. Tanzi and Schuknecht (1997), Tanzi (1999), Schneider (2005, 2012), Eilat and Zinnes (2002), Ahmuda et al. (2008) and Chaudhuri et al. (2006) shed light on shadow economic activities, its measurement and development across developed, developing and transition economies.

In reviewing recent literature, this paper focuses on a number of areas related to the shadow economy, in particular economic freedom, corruption and effective democracy and their relationship with economic growth. Researchers have identified several factors that are expected to have a negative association with the size of the shadow economy. These include trust (D'Hernoncourt and Meon 2012) and tax morale and quality of institutions (Torgler and Schneider 2009). Three key factors that have impacted the size of the shadow economy are: debt, default risk, corruption and financial development (Blackburn et al. 2012; Elgin and Uras 2013), information communication technologies (ICT) (Indjikian and Siegel 2005) and environmental violations (Biswas et al. 2012).

Considering economic freedom and corruption (Apergis et al. 2012), we particularly looked at economic developments and their variations (Pieroni and d'Agostino 2013), income and corruption (Saha and Gounder 2013), well-being (Belasen and Hafer 2012), benefits from economic freedom (Gehring 2013) and industrial policy and biased redistribution (Holcombe 2013). In some studies, there is evidence of a causality between economic freedom and economic growth (Narayan et al. 2011), competition law and health policy (Mossialos and Lear 2012), economic policies (Compton et al. 2011), capitalism as a necessary condition for economic freedom (Pryor 2010), transition from planned to market economies (Pääkkönen 2010) and inflows of FDI (Azman-Saini et al. 2010).

Considering an effective democracy, the focus in literature so far has been on the impacts of democracy on FDI (Asiedu and Lien 2011), the effectiveness of aid (Bjørnskov 2010) and benefits to political elites in recipient countries. Political culture (Feld and Kirchgässenr 2000) and press freedom (Kalenborn and Lessmann 2013) are found to affect the outcomes of economic policy. There is also an argument for degrowth policies (Johanisova and Wolf 2012) for facing multiple crises with environmental, social and economic dimensions and the different questions that this raises (Romano 2012). State capacity, democracy and rapid growth in industries (Rock 2009), and the importance of good institutional structures are important for economic performance (Knutsen 2013) and the effects of economic and political inequality on institutions (Savoia et al. 2010).

The rest of this study is organized into three major parts. The shadow economy and its measurement and magnitude across countries and country groups, factors determining its rate of increase and its linkages with environment and ICT are discussed in the first section. The second section is about economic freedom and corruption and their relationship with economic growth. The third section discusses effective democracy and how it impacts economic growth in nations under different resource endowments. The final section gives the conclusion.

2 Shadow Economic Activities

2.1 Shadow Economic Activities and Their Measurement

The shadow economy is a key source of the gap between observable and actual economic measures. It captures all the activities which are not measured by official activities and consists of both legal and illegal activities outside the reach of the government. It makes up a significant share of the overall economy around the world. Other synonyms for the shadow economy are hidden, black, underground, unobserved, unofficial, unrecorded and parallel economy. There is evidence that underground activities have been on the rise since the 1970s. This rise is attributed to the stronger presence of government activity in economies, increase in tax rates to finance larger public spending programs and in parallel the desire to escape taxes and regulatory restrictions (Tanzi and Schuknecht 1997). Tanzi (1999) suggests that the shadow economy is growing because of the presence of activities that are difficult to measure and tax.

Schneider (2005, 2012) considers shadow economic activities a fact of life. Most societies attempt to reduce the magnitude of the shadow economy by controlling activities through legal measures such as punishment and persecution or by preventive measures with investments in welfare and education. Despite significant investments in the collection of data on shadow economic activities, it is difficult to obtain accurate information about its nature and magnitude. Schneider talks of the existence of comprehensive literature on particular aspects of the shadow economy, but the subject remains controversial. Furthermore, there is disagreement among researchers about the definition and estimation procedures and their use in economic analyses and policymaking.

In a common approach, Schneider (2005) defines the shadow economy to include all market-based legal production of goods and services that are deliberately concealed from public authorities for the following reasons: (i) to avoid payment of income, value added or other taxes, (ii) to avoid payment of social security contributions, (iii) to avoid having to meet certain legal labor market standards, and (iv) to avoid complying with certain administrative procedures. However, this definition does not include economic activities that are illegal and fit the characteristics of classical crimes, as well as the informal household economy or tax evasion.

In another approach Eilat and Zinnes (2002) treat the shadow economy as a distinct entity, instead of seeing it just as a symptom of policy failures in transition countries. They examine its short-term and dynamic consequences for development. The shadow economy is measured in two different ways: first, the electricity method which attributes growth in total electricity consumption in excess of growth in GDP to the shadow economy, and the second a modified electricity approach correcting for limitations in the first approach.

However, in a third approach, Ahmuda et al. (2008) look at the monetary measure of the shadow economy. They discuss the money demand function and observed cash balances and their variations which are explained by variables which induce agents to make hidden transactions to estimate the size of the shadow economy. However, on econometric grounds, researchers have criticized the quantitative accuracy of this method. This critique is attributed to time series properties, structural breaks and sensitivity to units of measurement to lag the dependent variable and its initial condition.

2.2 Shadow Economic Activities in Industrialized and Transition Economies

Limited statistics from high income countries point to a positive trend in the development of activities in the shadow economy, yet little is known about its magnitude in transition, low-income and emerging economies. Schneider (2005) estimates the shadow economy for 110 countries (66 developing, 23 transition and 21 industrialized OECD) for 1990–1991, 1994–1995 and 1999–2000. The results provide some insights into the main causes and studies the dynamic effects of the shadow economy. The main causes of the shadow economy are found to be high tax and social security contribution burdens, the intensity of regulations and the low quality of public sector services.

Transition economies have undergone major changes. Increased unemployment, decline in GDP, a paralyzed bureaucracy and government corruption during this period saw a surge in the growth of shadow economic activities. Eilat and Zinnes (2002) conducted research on the shadow economy in transition countries. Their objective was to use a policy perspective to find out whether the shadow economy is a ‘friend’ or a ‘foe’. Their research was conducted in three parts: it lays out theoretical and empirical backgrounds, it estimates the size of the shadow economy and it examines its effects and discusses issues of policy implementation.

For the empirical part, Eilat and Zinnes (2002) measured the relative size of the shadow economy vis-à-vis the official GDP in 25 transition countries for 1990–1997. The patterns show that, once established, the shadow economy is hard to remove. Estimation results show that a dollar decline (rise) in official GDP is attenuated by a shadow economic expansion (contraction) of 31 (25) cents. Finally the authors examined whether the shadow economy prevents, slows down, or promotes economic growth and competitiveness, and through what mechanisms. In addition, they considered implications for policymaking that address key questions. The policy recommendations include actions with multiple benefits (foreign exchange management, better regulation, institutional strengthening, oversight, transparency and public participation, bank privatization and capital market development, decentralization and better local public finance, rule of law, further liberalization, macro-stabilization), actions that directly target the shadow economy (pay your bills, taxation, stricter, more strategic enforcement), actions whose effectiveness is changed by the shadow economy and implementation considerations (market exit, policy complementarity and coordination, selective targeting, dynamic considerations).

In the context of industrialized and transition economies, Schneider (2005), maintains that the shadow economy is expected to influence the tax system and its structure, the efficiency of resource allocation between sectors and the official economy in a dynamic sense. Therefore, several studies have integrated underground economies into macroeconomic models to facilitate an investigation of the effects of monetary and fiscal policies on formal and informal economies and economic growth. In the neoclassical view, the underground economy is assumed to provide the economy with a dynamic entrepreneurial spirit. It can lead to greater competition and higher efficiency, help create markets, increase financial resources, enhance entrepreneurship and transform the legal, social and economic institutions necessary for asset accumulation providing a higher potential for economic growth.

Schneider (2005) concludes that for all countries investigated, the shadow economy as share of GDP had reached a remarkably large size (Africa 33.9–41.2; Americas 34.2–41.5; Asia 20.9–26.3; transition countries 31.5–37.9 and highly developed OECD countries 13.2–16.8). The average percentage shares of GDP in all cases are increasing over time. The author demonstrates empirically a strong interaction of the shadow economy with government policies and with the official economy. He draws three further conclusions. First, an increasing burden of taxation and social security payments combined with rising state regulatory activities, are the major driving forces underlying the size and growth of the shadow economy. Second, the shadow economy has a statistically significant and quantitatively important influence on the growth of the official economy. Increases in the shadow economy have a negative effect on official growth in a developing country, but a positive effect in developed industrialized and transition countries. Finally, shadow economies are a complex phenomenon, and are present in all types of economies. People engage in shadow economic activities because of government actions, most notably high levels of taxation and regulation.

Using state level data from India, Chaudhuri et al. (2006), investigate the size of the hidden economy in Indian states during 1974–1975 and 1995–1996. Their results show evidence of a varying size of the hidden economy in Indian states (10.2 till 48.3 % of GDP), and this provides evidence in favor of liberalization of the Indian economy in 1991–1992, which reduced the growth of the hidden economy. In addition, the results show that growth in the size of the shadow economy is less in election years, lower in states governed by coalitions of political parties and that increased growth of newspapers and literacy rates translate into cleaner governance and help lower shadow economy activities.

2.3 Factors with Negative Association with Shadow Economy

There are several factors that are expected to have a negative association with the size of the shadow economy. These include trust, tax morale and quality of institutions. This section briefly discusses these factors considering the magnitude and development of shadow economic activities.

Explaining the shadow economy, some researchers go beyond the objective variables such as tax burden, rate of public expenditure or the density of regulation, and use subjective variables such as perceptions, expectations, attitudes and motivations such as tax morale or institutional quality. The relationship between tax morale and institutional quality and the shadow economy is investigated by Torgler and Schneider (2009). The shadow economy is measured as a percentage of the official GDP using the DYMIMIC-method where one estimates the parameters for determining the size of the shadow economy. European Values Survey data on cheating on taxes and Latinobarometro data on the justifiability of avoiding paying taxes are measures of tax morale, and the quality of governance index is used as a proxy for institutional quality. These allow an investigation of the impact of such factors at the macro level. Torgler and Schneider use a multivariate analysis to examine the quantitative impact of these factors on the level of and changes in the shadow economy. After controlling for a variety of potential factors, they find strong support for the hypothesis that higher tax morale and higher institutional quality lead to a smaller shadow economy. The results show that improving social institutions (enhancing tax morale, voice and accountability, the rule of law, government effectiveness and its regulatory quality and reducing corruption) helps reduce the incentive to 'go underground'. In addition, the legal structure and security of property rights are among the important factors that influence the size of the shadow economy.

Another factor with a significant potential impact on the size of the shadow economy is trust. Trust can be a substitute for formal and legal contracts in a situation when the agents involved in shadow transactions cannot rely on the formal legal system to enforce agreements or settle disputes. This view suggests that trust increases the size of the formal sector by negatively impacting the size of the informal sector. In this regard, D'Hernoncourt and Meon (2012) investigate the relationship between trust and the size of the shadow economy. They report a negative relationship between the size of the shadow economy and generalized trust. Their data includes 145 developed and developing countries that they observed over 1999–2003. They define trust as a view that most people can be trusted which is measured as the trust index provided in the WVS data. Comprehensive sensitivity analyses that they conducted confirm that the relationship is robust to controlling for various sets of economic, policy and institutional variables, to changing the estimate of the shadow economy and the estimation period, as well as to controlling for endogeneity. Trust and the shadow economy are negatively related and trust matters more for developing countries. The impact runs through agents' propensity to shy away from paying taxes. However, the shadow economy is found to be independent of trust in institutions and from income inequalities. It is mainly present in developing countries. The findings suggest that the tax compliance effect of trust dominates its role as a substitute for the legal system.

2.4 *Factors with Positive Association with Shadow Economy*

The extent, use and abuse of three key factors have impacted the size of the shadow economy: corruption, ICT and environmental violations. Water quality and air pollution are serious problems in many developing countries threatening the health of citizens. Many of the environmental problems are fostered by the sizeable shadow economies in developing and transition economies. Many countries lack regulations or their governments are corrupt or unable to implement effective environmental policies.

Biswas et al. (2012) are surprised that there is still lack of research on the shadow economy-environment nexus. They fill the gap in literature by testing the extent to which the informal sector contributes to pollution and the extent to which corruption undermines environmental policy. Their empirical study depending on the model and the sample size varies between 107 and 134 countries for the period 1999–2005. Two hypotheses are tested: the shadow economy increases pollution, and corruption exacerbates the effect of the shadow economy on environmental degradation. Pollution is measured as per capita environmental pollution. The shadow economy is measured as a share of the GDP and to measure corruption, the corruption index from Political Risk Services is used. They find that the larger the shadow economy, the greater the pollution, but the effect declines with controlling corruption. The results also hold when controlling for economic and demographic covariates. They also take into account country group heterogeneity by level of income. Their findings imply that policymakers may address the effects of corruption and the shadow economy before increasing environmental standards and regulations.

The diffusion and spread of internet technology in recent decades has had consequences for different sectors of the economy. The effects span from business to banking, e-government, education, information, productivity, foreign direct investment, inflation and political economy issues. In some cases, researchers point to its positive impacts on economic outcomes such as inflation reduction, higher volume of trade, increased productivity and higher economic growth. Indjikian and Siegel (2005) provide a survey of the economic effects of the spread of ICT while Elgin (2013) asks whether the spread of the internet has aided or abetted the shadow economy. The theoretical framework argues that internet usage creates two effects on the size of the informal sector through increasing productivity and reducing the size of the shadow economy and tax evasion. Using a panel data of 152 countries over 9 years (1999–2007), Elgin examines the empirical relationship between the degree of internet usage and the size of the shadow economy. Informal sector size, total factor productivity (TFP) growth and taxes are estimated as a system where internet usage explains TFP growth and taxes which in turn determine the size of the informal economy accounting for various control variables. The estimation results indicate that the association between internet usage and the size of the shadow economy strongly interacts with GDP per capita and level of development. In designing an economic policy for the shadow economy, it is recommended that policymakers should take into account its two opposite effects on productivity and tax aversion. Policy measures which put the productivity enhancing effect forward relative to the tax-evasion effect should be taken. Subsidizing ICT investment and better infrastructure, improving institutions and bureaucratic quality are among the steps that the governments need to take.

2.5 Public Finances and the Size of the Informal Economy

Data on sovereign debt and the size of the shadow economy vary substantially among countries. The macroeconomic implications include high interest rate payments which constraint economic well-being, risk of debt default, limited enforcement of tax collections and the presence of the informal sector, which influences fiscal instruments, tax revenues and the government's ability to repay public debt. Elgin and Uras (2013) address the interactions between government's indebtedness, sovereign default risk and the size of the informal sector. They test a theory which suggests that in societies with limited tax enforcement, the presence of informality constrains the set of fiscal policy alternatives. It increases public debt and the implied probability of sovereign debt restructuring. In the empirical part, a number of hypotheses are tested: a larger size of the informal sector is associated with higher public indebtedness, higher interest rates paid on sovereign debt, a higher level of financial instability and a higher probability of sovereign default. These are represented by the ratio of public debt to GDP, the interest rate spread, the financial stress index (FSI) and default probability obtained from the World Development Indicators (WDI) database.

The empirical results in Elgin and Uras (2013) from cross-country panel regressions of 152 countries over the period 1999–2007 show that after controlling for previously highlighted variables identified in literature that could explain variations in financial instability, sovereign default risk and public indebtedness, the size of informality in the economy remains a significant determinant of these variables. In order to show that a larger shadow economy size is associated with the four possible undesirable outcomes, they compute elasticities. A 1 % increase in the size of the informal sector led to a 6 % increase in the ratio of public debt to GDP, a 127 % increase in the financial stress index led to a 14 % increase in the interest rate spread and a 4 % increase in the probability of default. They conclude that improving enforcement through enhancing law and order and implementing policies for lower taxes and unemployment can have significant effects on the sovereign default risk. Policy implications suggest that public policy should focus more on taking measures to reduce the size of the shadow economy. Increasing law enforcement, enhancing institutional development, policies towards lowering income taxes and unemployment are among the prescribed measures recommended for policymakers in societies with large informal sector sizes.

In a recent study the theoretical relationship between the underground economy and financial development in a model of tax evasion and bank intermediation is studied by Blackburn et al. (2012). Here the presence of an underground sector is a reflection of individuals' incentives to conceal their economic activities or circumstances for two reasons. First, these activities would be less rewarding if practiced in the formal sector. Second, because the activities are illegal. The shadow economy has potentially serious implications for economic performance and public policy. The primary objective of this study was to shed light on the determinants of underground activities and their influence on an individual's incentives. The results show that the lower the financial development, the higher the incidence of tax evasion and the greater the size of the underground economy. This negative relationship implies that the marginal net benefit of income disclosures increases with the level of financial development.

3 Economic Freedom, Corruption and Economic Growth

3.1 Corruption and Economic Freedom

A few studies discuss the substitutability of corruption and economic freedom. Their aim is to analyse how economic freedom affects corruption. However, an analysis of the relation between corruption and economic growth is of great interest. It is expected that the more the government intervenes in markets, the more the economic freedom declines. Apergis et al. (2012) investigate the relationship between economic freedom and corruption using data from 50 US states during 1981–2004. Their study contributes to literature in a number of respects. First, it uses a more objective measure of corruption, namely, the number of government officials convicted in a state for crimes related to corruption. Second, it exploits both time series and cross-sectional variations in the data. Three different measures of economic freedom are used. The results show that in the long-run, economic freedom, per capita income and education have a negative and statistically significant impact on corruption whereas income inequality has a positive impact. The causality tests reveal bidirectional causality between economic freedom and corruption in both the short and long-run.

It is expected that economic freedom reduces levels of corruption. Empirical studies support this expectation, but with some inconsistencies in results. Therefore, the prediction that economic freedom is beneficial in reducing corruption is not found to be universally robust in empirical studies. Pieroni and d'Agostino (2013) reviewed this relationship by using firms' data in a cross-country survey in 1999.

They argue that approaches using aggregated macro data have not been able to explain this appropriately. They model cross-country variations of the economic freedom-corruption relationship using multilevel models. They also disentangle the determinants for several components of economic freedom with unequal effects. The results show that the extent of macro-effects on measures of microeconomic freedom for corruption can explain why lack of competition policies and government regulations may yield more corruption but with large variability across countries. Estimates for Africa and transition economy sub-samples confirm these results.

It is well-known that corruption is detrimental to economic performance, especially in developing countries. However, it is unclear whether an increase in income consistently reduces corruption. Saha and Gounder (2013) investigate the relationship between income and corruption to provide an insight into changes in the level of corruption and economic development. For the empirical part they use data from 100 countries classified by region and income for the period 1995–2008. Unlike most previous studies, they use hierarchical polynomial regression to evaluate the existence of non-linear relationships after controlling for socioeconomic and institutional factors. They find a quadratic function to best fit the data. The results show that corruption is explained by various socioeconomic, political and institutional factors. The results also challenge some of the findings of a negative income-corruption association. Despite an upsurge in corruption among the low-to-medium income countries, advanced stages of development eventually reduce corruption levels. Policy implications suggest that a combination of economic, institutional and social policies can effectively reduce and lower the effects of corruption. Achieving a high level of educational attainment, employment opportunities and equal income distribution are necessary for discouraging corrupt activities.

3.2 Gains and Well-Being from Economic Freedom

Interest in factors that increase well-being, happiness and life satisfaction has increased over past decades. Belasen and Hafer (2012) state that there is plenty of evidence that well-being is positively related to the level of general intelligence and economic freedom across countries. Their aim is to determine whether economic freedom and well-being are related at the state level. Well-being is measured as a multidimensional index and economic freedom as an index of freedom. A regression analysis of the link between well-being and economic freedom accounting for different control variables indicates that across the 50 US states, improvements in economic freedom lead to higher levels of well-being. They find that the relationship between well-being and economic freedom differs significantly across regions in the US and they suggest that the government plays a major role in influencing individual well-being.

In relation to the link between economic freedom and well-being, Gehring (2013) asks the question: who benefits from economic freedom? His focus is on the effects of economic freedom on subjective well-being. Well-being is defined as happiness obtained from the World Value and the European Value surveys. Results from a panel of 86 countries over the 1990–2005 period suggest that economic freedom has a significant positive effect on subjective well-being. Economic freedom's dimensions including legal security and property rights, sound money and regulation are strong predictors of higher well-being. Different sensitivity analysis tests show that the positive effect is not affected by socio-demographic variables. Developing countries profit more from higher economic freedom by reducing the regulatory burden which strengthens the effect. Culture seems to moderate the effect. Gehring concludes that societies that are more tolerant and have a positive attitude towards the market economy profit the most in the form of well-being.

The issue of inequality in the distribution of freedom, growth and well-being is an important one. For instance, many South Koreans believe that the country's remarkable economic successes are the product of an industrial policy that made Korean industry competitive. Holcombe (2013) maintains that while there is widespread sentiment in favor of maintaining the emphasis on industrial policy, some also believe that the industrial policy has generated unequal benefits at the expense of the welfare of the working class. In recent years, there have been tendencies towards economic democracy to share gains from increased productivity more equally. Neither industrial policy nor economic democracy is in the best interest of the Koreans as each rewards particular interest groups. A laissez faire policy of minimal government interference is suggested to provide the best environment to foster South Korea's continued economic progress.

3.3 *Causality Between Economic Freedom and Economic Growth*

Causality and its direction between economic freedom and economic growth has been another focus of attention of the kind of cross-national comparative literature under review here. Narayan et al. (2011) examined the relationship between democracy and economic growth in 30 Sub-Saharan African countries for the period 1972–2001. Democracy is defined as democracy index constructed by Freedom House and the Legislative Index of Electoral Competitiveness. The results support the Lipset hypothesis indicating that in the long-run, real GDP Granger causes[1] democracy and an increase in GDP results in an improvement in democracy in a number of countries. Support for the compatibility hypothesis that in the long run democracy Granger causes real income and an increase in democracy has a positive effect on real incomes is also found in a number of countries. Support for

[1] A time series X is said to Granger-cause Y if it can be shown through a series of statistical tests on lagged values of X that the X values provide significant information about future values of Y.

the conflicting hypothesis that in the long-run democracy Granger causes real income and an increase in democracy has a negative effect on real income is found for two countries. For most countries there is long-run Granger neutrality between democracy and real GDP supporting the skeptical hypothesis for a majority of the countries.

European community (EC) welfare states have developed compulsory public social insurance to promote social cohesion among citizens. The European health systems are based on the principles of solidarity, equity and universality. In recent years the EC health policy shows signs of tension between EC economic freedom and social policy. The EC competition law restricts national health policy options of member states and limits the ability to adequately account for non-economic gains in consumer welfare. Mossialos and Lear (2012) define the policy and legal parameters of the debate between competition law and health policy in EC. They use a sample of cases to analyse how the European Court of Justice, national courts and national competition authorities have applied competition laws to the health services sector. The health policy does not enjoy special exclusion from competition law which above all emphasizes trade. They conclude that the implications of the convergence of recent trends in competition law enforcement and health system market reforms do not undermine the social model and its goals of equity and social cohesion.

Compton et al. (2011) analyse which economic policies are most favorable for economic growth. In literature, policies that promote economic freedom have been suggested as a viable path towards sustained economic growth. Using the multidimensional measures of the economic freedom index the authors investigate the nature of the relationship between economic freedom, the size of the government, taxes, labor market and economic growth for the 50 US states spanning the period 1981–2004. The results from a comprehensive sensitivity analysis with respect to variable definitions and model specifications as well as accounting for endogeneity and a selection bias and after controlling for various control variables, show evidence of a significant positive relationship between economic freedom and economic growth at the state level. However, not all selected components of economic freedom affect growth equally. As policy recommendations, constraining excessive government expenditure, minimizing the tax burden and maintaining an open labor market are suggested.

3.4 Economic Transition, Freedom and Growth

Over the last two centuries the spread of capitalism and the level of political freedom have increased. The connection between economics and politics has made economic freedom indispensable for achieving political freedom. Pryor (2010) tests Milton Friedman's conjecture that capitalism is a necessary condition for political freedom. An analysis of data shows that for the decade around 2000, indices of the degree of capitalism and the degree of political freedom with few exceptions were highly correlated. This provides plausibility for Friedman's conjecture. However, results based on data for the nineteenth century refute Friedman's conjecture. These contradictory results are related to per capita GDP and the educational levels of the population. Despite their correlations, capitalism is neither a necessary nor a sufficient condition for political freedom and no confirmation is found for the reverse proposition.

The 1991 collapse of the Soviet Union led to the creation of 15 new states and a transition from a centrally planned economy to a market economy for 25 states for the period 1998–2005. These countries experienced heterogeneous growth and development. Pääkkönen (2010) reviews the political economy of economic growth in post-communist economies' transition to free markets. His focus is on the role of economic policy and institutions in the transition process. The author tests the hypothesis that better institutions, measured in terms of economic freedom, contribute to growth. The empirical results confirm this hypothesis. There is an interactive effect between economic freedom and investment, where institutions affect growth indirectly by affecting the marginal effects of investment. He concludes that non-linearities are present in the specified growth model, where the marginal effect turns to become negative. Increased government consumption also has a negative effect on growth suggesting the presence of wasteful spending and a hindrance to growth.

The relationship between foreign direct investment (FDI) and its technology, management and skill spillovers and improved productivity and economic growth is intensively analysed with conflicting conclusions. Efforts have been made to identify determinants of FDI inflows to design suitable policies and to remove restrictions imposed on foreign capital flows. Azman-Saini et al. (2010) investigate the systemic link between economic freedom as absorptive capacity, FDI and economic growth in a panel of 85 countries for the period 1976–2004. Their empirical results based on the system estimator, reveal that FDI has no direct positive effect on output growth. Instead, the effect of FDI is contingent on the level of economic freedom in the host countries. Countries promoting greater freedom of economic activities gain from the presence of multinational corporations. The results suggest that economic freedom is important for host countries' absorptive capacities. Policymakers should promote better freedom of economic activities.

4 Effective Democracy and Economic Growth

4.1 Democracy, Natural Resources and Aid

In recent decades, attitudes towards FDI have changed. Researchers and international organizations encourage developing countries to attract FDI as a tool for fighting poverty. However, most developing countries have weak democracies.

Asiedu and Lien (2011) examine the impact of democracy on FDI assuming that the relationship between the two is the same for resource exporting and non-resource exporting countries. Their objective is to examine whether natural resources in host countries alter the relationship between democracy and FDI. Results based on data from 112 developing countries over the period 1982–2007 show that democracy promotes FDI. However, this effect is observed only if the share of natural resources in total exports is less than some critical value. Of the sample, the authors identified 90 countries where an expansion of democracy may enhance FDI inflows, while in 22 countries it may reduce FDI. The effect of democracy on FDI is found to depend on the size and not the type of natural resources exported. It is also well-known that in resource rich countries democracy is weak and the curse of natural resources leads to lower economic growth.

Literature on the measurement of the effectiveness of aid is voluminous. Foreign aid is found to be ineffective in creating economic growth. Bjørnskov (2010) refers to a popular argument for the absence of beneficial effects—that it is skimmed by political elites in recipient countries. Some studies also suggest that aid may be more effective in relatively democratic developing countries. He studied data for 88 developing countries for the period 1960–2000 with a 5-year average to test the relationship between democracy and the effectiveness of aid. The results indicate that foreign aid and democracy in conjunction are associated with a higher share of income held by the upper quintile as a result of rent seeking during the transition to democracy. The results also show that foreign aid leads to more skewed income distribution in democratic developing countries while the effects are negligible in autocratic countries. The paper also has a discussion on potential mechanisms generating this perverse effect and questions whether equitable income distribution and democratization can be achieved through foreign aid.

4.2 Political Culture, Press Freedom and Economic Policy

Feld and Kirchgässenr (2000) in their research show that the political culture in Switzerland is influenced by the country's system of direct democracy. To support their statement, they consider the outcomes of two political decisions by Swiss citizens in the 1989 vote on an initiative that proposed to abolish the Swiss army and the 1979 vote on a possible exit from the nuclear energy system. The outcome in both the cases was unexpected. The two cases showed that compared to purely representative systems, direct democracy led to a different type of communication among citizens and also between citizens and their representatives. The results show that the opportunity of deciding on political issues by themselves provides citizens with greater incentives to collect more information. Better informed citizens and a focus on common interests leave less room for politicians to pursue their personal interests. As a consequence of better informed citizens who enjoy and apply direct democratic rights, public expenditure and public debt are lower. Increased participation in decision making and the feeling of being responsible for their community has resulted in lower tax evasion compared to representative democratic systems with less participation.

In particular, corruption is widespread in developing countries. A precondition for achieving growth and reducing poverty is dealing with corruption in an effective way. This view is also supported by growing literature. Potentially effective instruments for rooting out corruption are democratic elections and press freedom. Since their effects on corruption interdepend on each other a simultaneous interactive model is used to estimate the conditional effects. Kalenborn and Lessmann (2013) empirically analyse the joint impact of democracy and press freedom on corruption. They argue that both institutional features complement rather than substitute each other in controlling corruption. Corruption is proxied by the Transparency International's Corruption Perception Index, democracy is represented by Vanhanen's Democratization Index and press freedom is provided by Freedom House. Result estimates based on a cross-section of 170 countries covering the period 2005–2010 as well as based on a panel evidence for 175 countries from 1996 to 2010 show that democratic elections only work in controlling corruption if there is a certain degree of press freedom in a country, and vice versa. The policy implications of the results are that democratic reforms are more effective if they are accompanied by institutional reforms strengthening the monitoring of rent seeking representatives.

4.3 Political Democracy and Degrowth Policies

Johanisova and Wolf (2012) state that we are living in times of quickly shifting paradigms where ideals like economic growth and political democracy are falling and losing their appeal in the face of multiple crises with environmental, social and economic dimensions. In contrast to political democracy with power control in the public sector, economic democracy has received much less attention. The authors define economic democracy as 'a system of checks and balances on economic power and support for the right of citizens to actively participate in the economy regardless of social status, race, gender, etc.' They suggest six possible faces of economic democracy and look at their implications for the vision of a sustainable, equitable and non-growing society including regulation of market mechanisms and corporate activities to curb economic power; support for social enterprises which are better suited to a degrowth economy; democratic money creation processes; and pluralist community currencies. The authors see reclaiming the commons as an important aspect of enhancing economic democracy; redistribution of income and capital assets is another approach to achieving economic democracy; and finally, a broader view of economic democracy will involve a diversity of production scales and modes. Implementation of these visions will mitigate growth pressures from profit-oriented enterprises and facilitate easier implementation of degrowth policies.

In relation to the link between democracy and degrowth, Romano (2012) elaborates on four questions: can the degrowth project contribute to rediscovering the meaning of democracy? Can the establishment of a real democracy lead to building a degrowth society? Has the project of a 'democratic degrowth' a real chance to succeed, from a political point of view? If not, how should degrowth be re-thought to foster democracy? Romano's belief is that the currently dominant formulation of degrowth and its relationship to democracy are highly problematic. A degrowth proposal reproduces a central paradox where degrowth does not aspire to restore collective linkages. It only works like an intervention offering techniques that will allow the human species merely to stay alive. Moreover, there is a political problem: while the need for degrowth is presented as very urgent, the tactic for spreading the idea is one of an elitist strategy of voluntary simplicity, which can only work very slowly. As a result, degrowth cannot connect to real social processes covering a large part of the population. To make degrowth feasible and to restore democracy, Romano proposes a foundation of degrowth on a human subject of 'de-thinking' and discusses the political pathways for such a degrowth project.

4.4 State Capacity and Economic Growth

In general, growth in Asia's authoritarian regimes has been very high. This gives rise to the question: has democracy and/or democratization slowed growth in Asia? Rock (2009) tests hypotheses linking Asia's democracies and autocracies to growth. The tests are conducted in a panel regression framework that controls for fixed country and time effects, other major variables affecting growth and for endogeneity between explanatory variables resulting from the direct effect of democracy and its indirect effect through investment. The '*democracy slows growth*' hypothesis is tested against the bureaucratically capable authoritarian regimes of East Asia's developmentally minded governments. The findings from various model specifications and sensitivity analyses reject the '*democracy slows growth*' hypothesis. They show that unlike autocracy, democracy by itself causes growth and investment to increase in Asia. The implication of these findings is that concern over the economic growth performance of new democracies in Asia is misplaced.

It is well-known that good institutions are important for economic performance. However, little is known about the effects of specific institutional structures. Knutsen (2013) discusses how regime type and state capacity may interact in jointly affecting economic growth. A number of hypotheses are drawn from the trade-off between state capacity and democratization in enhancing economic growth. State capacity is measured as a bureaucracy quality index and other control variables include regime duration, the ethnic fractionalization index, plurality, religion and data on the colonizer (if any). Knutsen's analysis finds a positive and robust effect of democracy on growth in Sub-Saharan Africa, a continent characterized by weak-capacity states. The author identifies a robust interaction effect between democracy and state capacity on growth, both in Africa and globally. The effect of democracy on growth increases when state capacity is reduced. Democracy has a positive effect on growth in weak-capacity states, but not in high-capacity states. The results also indicate that state capacity enhances growth only in dictatorships confirming a trade-off effect between state capacity and democracy in affecting growth. The robustness of the results is confirmed through a sensitivity analysis.

Institutions play a major role as determinants of growth and development. Political economy factors such as inequality and political democracy influence the quality of institutions. Savoia et al. (2010) survey theoretical and empirical literature on the effect of economic and political inequalities on institutions. The current understanding suggests that unequal societies develop exploitative and inefficient institutions. The most used indicators in empirical research include property rights, regulatory quality, government effectiveness, political stability, executive constraints, democracy index, income inequality index and land inequality. Empirical research has mainly concentrated on the cross-national level and supports, to some extent, the existence of an inverse relationship between inequality and institutions. The authors suggest further research but based on country, state and micro-level analyses as complements to aggregate level studies. They also recommend an investigation of the separation of the direct and indirect effects of inequality on institutions through democratization.

5 Concluding Remarks on the Dark Side of Economics

The shadow economy and corruption are the two main dark sides of economics. Activities in the shadow economy have been on a rise around the world regardless of the level of development, geographic location or type of governance in the country. This has led to significant violations of laws and regulations and has reduced tax revenue collections, caused statistical discrepancies, inequalities, corruption and increased public budget deficits and subsequent public debt default problems. This paper reviewed literature on the shadow economy and its direct and indirect effects on economic growth. Several studies attribute the shadow economy to higher taxes, larger governments and their interventions, restricted economic freedom and the desire to escape taxes and regulatory restrictions by individuals and businesses.

This paper reviewed a large number of recent studies investigating theoretical and empirical aspects of the shadow economy, in particular the measurement of the shadow economy and its causality with economic growth. In the process, emphasis was placed on issues such as corruption, economic freedom, state and institutions' capacities and quality as well as on morale and trust. For the empirical parts, these studies use large datasets allowing for cross-country and country group comparisons. A negative relationship is found between shadow economic activity and level of development, democracy and press freedom. Countries with abundant natural resources for exports are facing a larger degree of activities in the shadow economy, corruption, inefficiency, inequality and lower economic growth. In literature this is referred to as the 'curse of natural resources'.

References

Ahmuda H, Alvaredo F, Canavese A (2008) The monetary method to measure the shadow economy: the forgotten problem of the initial conditions. Econ Lett 101:97–99

Apergis N, Dincer OC, Payne JE (2012) Live free or bribe: on the causal dynamics between economic freedom and corruption in U.S. states. Eur J Polit Econ 28:215–226

Asiedu E, Lien D (2011) Democracy, foreign direct investment and natural resources. J Int Econ 84:99–111

Azman-Saini WNW, Baharumshah AZ, Law SH (2010) Foreign direct investment, economic freedom and economic growth. Econ Model 27:1079–1089

Belasen AR, Hafer RW (2012) Well-being and economic freedom: evidence from the States. Intell 40:306–316

Biswas AK, Farzanegan MR, Thum M (2012) Pollution shadow economy and corruption-theory and evidence. Ecol Econ 75:114–125

Bjørnskov C (2010) Do elites benefit from democracy and foreign aid in developing countries? J Dev Econ 92:115–124

Blackburn K, Bose N, Capasso S (2012) Tax evasion, the underground economy and financial development. J Econ Behav Organ 83:243–253

Chaudhuri K, Scheneider F, Chattopadhyay S (2006) The size and development of the shadow economy-an empirical investigation from states of India. J Dev Econ 80:428–443

Compton RA, Giedeman DC, Hoover GA (2011) Panel evidence on economic freedom and growth in the United States. Eur J Polit Econ 27:423–435

D'Hernoncourt D, Meon PG (2012) The not so dark side of trust: does trust increase the size of the shadow economy? J Econ Behav Organ 81:97–121

Eilat Y, Zinnes C (2002) The shadow economy in transition countries: friend or foe? A policy perspective. World Dev 30(7):1233–1254

Elgin C (2013) Internet usage and the shadow economy Evidence from panel data. Econ Syst 37 (1):111–121

Elgin C, Uras BR (2013) Public debt, sovereign default risk and shadow economy. J Financ Stabil 9(4):628–640

Feld LP, Kirchgässenr G (2000) Direct democracy, political culture, and the outcome of economic policy: a report on the Swiss experience. Eur J Polit Econ 16:287–306

Gehring K (2013) Who benefits from economic freedom? Unraveling the effect of economic freedom on subjective well-being. World Dev 50:74–90

Holcombe RG (2013) South Korea's economic future: industrial policy or economic democracy? J Econ Behav Organ 88:3–13

Indjikian R, Siegel DS (2005) The impact of investment in IT on economic performance: implications for developing countries. World Dev 33:681–700

Johanisova N, Wolf S (2012) Economic democracy: a path for the future. Futures 44:562–570

Kalenborn C, Lessmann C (2013) The impact of democracy and press freedom on corruption: conditionality matters. J Pol Model 35(6):857–886

Knutsen CH (2013) Democracy, state capacity, and economic growth. World Dev 43(1):1–18

Mossialos E, Lear J (2012) Balancing economic freedom against social policy principles: EC competition law and national health systems. Health Policy 106:127–137

Narayan PK, Nayaran S, Smyth R (2011) Does democracy facilitate economic growth or does economic growth facilitate democracy? An empirical study of Sub-Saharan Africa. Econ Model 28:900–910

Pääkkönen J (2010) Economic freedom as driver of growth in transition. Econ Syst 34:469–479
Pieroni L, d'Agostino G (2013) Corruption and the effects of economic freedom. Eur J Polit Econ 29:54–72
Pryor FL (2010) Capitalism and freedom? Econ Syst 34:91–104
Rock MT (2009) Had democracy slowed growth in Asia? World Dev 37(5):941–952
Romano O (2012) How to rebuild democracy, re-thinking degrowth. Futures 44:582–589
Saha S, Gounder R (2013) Corruption and economic development nexus: variations across income levels in a non-linear framework. Econ Model 31:70–79
Savoia A, Easaw J, McKay A (2010) Inequality, democracy, and institutions: a critical review of recent research. World Dev 38(2):142–154
Schneider F (2005) Shadow economies around the world: what do we really know? Eur J Polit Econ 21:598–642
Schneider F (2012) The shadow economy and work in the shadow: what do we (not) know? IZA Discussion Paper 2012: 6423
Tanzi V (1999) Uses and abuses of estimates of the underground economy. Econ J 109:338–340
Tanzi V, Schuknecht L (1997) Reconsidering the fiscal role of government: the international perspective. Am Econ Rev 87:164–168
Tausch A, Heshmati A, Karoui H (2014) The political algebra of global value change: general models and implications for the Muslim world. Nova Science, Hauppauge, NY
Torgler B, Schneider F (2009) The impact of tax morale and institutional quality on the shadow economy. J Econ Psychol 30:228–245

Part IV
Inclusive and Sustainable Growth

Antecedents of Environmentally Friendly Manufacturing Practices Among SMEs in Africa: Evidence from Uganda

Dan Ayebale, Esther Nafunka, and Ahurra Hope Ayebale

Abstract Environmental management is increasingly becoming an important topic of discussion in the business world today. Stakeholders as well as policymakers are demanding more accountability from companies in relation to their effects on the environment. In fact, putting the environment at the heart of a company's marketing drive has become a popular strategy as companies search for ways to achieve competitive advantages in the currently dynamic business landscape. This paper addresses these issues in a rarely studied context. Specifically, it documents empirical evidence on the nature of small and medium-sized enterprises (SMEs) adopting environmentally friendly manufacturing practices in a developing-country context where firms have a weak resource base and operate in a poor regulatory regime. By focusing on SMEs as opposed to the conventional focus on large corporations, and using a developing-country context, this paper attempts to contribute to extant literature by uncovering additional facets of the current topic with potential significant implications for business practice and public policy.

Keywords Environmental management • Recycling • Waste disposal • Small and medium-sized enterprises • Developing-country

---

D. Ayebale (✉)
Faculty of Business and Administration, Uganda Christian University, Box 4, Mukono, Uganda

Uganda Technology and Management University, School of Business and Management, Box 73307, Kampala, Uganda
e-mail: dayebale@ucu.ac.ug

E. Nafunka
Faculty of Business and Administration, Uganda Christian University, Box 4, Mukono, Uganda
e-mail: queegeo@yahoo.com

A.H. Ayebale
Faculty of Education, Department of Teacher Education and Development Studies, Kyambogo University, Box 1, Kampala, Uganda
e-mail: ayebalehope@yahoo.com

A. Heshmati (ed.), *Economic Integration, Currency Union, and Sustainable and Inclusive Growth in East Africa*, Advances in African Economic, Social and Political Development, DOI 10.1007/978-3-319-30432-8_8

1 Introduction

Environmental management is increasingly becoming an important topic of discussion in the business world today (Gadenne et al. 2009; Lefebvre et al. 2003; Williamson et al. 2006). Stakeholders as well as policymakers are demanding more accountability from companies in relation to their effects on the environment (Gadenne et al. 2009; Oba and Fobio 2012; Revell et al. 2008; Williamson et al. 2006). Putting the environment at the heart of a company's marketing drive has even become a popular strategy as companies strive to achieve and sustain competitive advantages in the highly dynamic global business landscape (Brammer et al. 2012; Simpson et al. 2004). In the conceptual domain, advances have been made to illustrate the importance of incorporating environmental issues in a company's strategy. For instance, according to Hamdouch and Depret (2012) a company can obtain good market evaluation following its attempts to address current environmental challenges in the production process. In some circles, a company's environmental attitude has been linked to its financial performance. For instance, the Global Environmental Management Initiative (GEMI) (2004) argues that a company's prowess in innovating in its products while taking the environment into consideration can enhance shareholders' value. Beyond the financial benefits, the Organization for Economic Cooperation and Development (OECD) (2007) and the United Nations Conference on Environment and Development (1992) also anticipate the benefits of reputation that may arise from developing a business strategy with the goal of reducing a firm's effects on the environment.

But not withstanding the conceptual propositions made in this area, there is a dearth of empirical research addressing the importance of building an environmentally responsive business especially from the context of SMEs (Brammer et al. 2012; Ki-Hoon 2009). The scant empirical research ascertaining the progress made by companies in building environmentally sustainable businesses has majorly focused on developed countries (Berry and Randinelli 1998). Empirical research focusing on developing countries especially from a SME's perspective in Africa is virtually non-existent. In line with the existing concerns about the relevance of findings arrived at in the context of developed countries to inform managerial practices in developing countries, this study seeks to explore the nature of environmental management practices among SMEs and the associated factors drawing on the context of Uganda.

Uganda is one African country that has registered tremendous progress in the arena of business development. It is currently one of the leading entrepreneurial economies on the continent. Its government has come up with a number of regulations to enhance SME response to current environmental challenges in the country (for example, the National Environmental Action Plan of 1994, National Environment Act of 2006 and the National Environmental Management Authority). In addition, firms in Uganda operate under a highly competitive environment of limited or no government protection. The liberal economic environment as well as the government's preference for open external competition has created a business milieu of numerous actors in the market including local and foreign firms. In this environment finding means to conserve resources has become a necessary practice for SMEs if they are to survive in the local competitive landscape. Focusing on Uganda can therefore offer some valuable insights into how local firms in a unique setting in Africa are responding to the general call around the world to integrate environmental sustainability in business operations. In the same vein, this study promises to contribute towards our understanding of the creative ways through which SMEs are positively responding to current environmental concerns by drawing on the context of one of the least-developed but highly dynamic local economies in Africa.

2 Literature Review

A rich body of research exists that provides insights into a number of issues which firms face in the adoption of environmentally friendly manufacturing practices (Brammer et al. 2012; Gutowski et al. 2005; Seidel et al. 2009). Key concerns in this literature are the barriers to adoption in terms of lack of information about manufacturing best practices (Biondi et al. 1997; Herren and Hadley 2010; Okello-Obura et al. 2007; Seidel et al. 2008; Weesasiri and Zhang 2012); limitations of firms to ask the necessary questions which are vital for adopting environmentally friendly manufacturing practices (Herren and Hadley 2010; Walker et al. 2008) as well as limited technical competence to incorporate environmentally friendly manufacturing practices (Seidel et al. 2009). From a strategic perspective, a debate has arisen regarding whether or not the adoption of environmentally friendly manufacturing is a core activity which provides a competitive advantage to firms (Avram and Kuhne 2008; Herren and Hadley 2010). While some researchers hold the view that investments in such activities may constrain firms to adequately execute their core activities, in more recent works, firm managers are encouraged to increase their investments in environmentally friendly manufacturing practices as a means of attaining strategic advantages such as reduction in raw material inefficiency; cost reduction and savings; regulatory compliance; an improved company image among the local community and stakeholders; higher employee commitment; improved product quality and public relations (Gadenne et al. 2009; Simpson et al. 2004).

Further, research has also made vital advances in relation to the process of successful implementation of environmentally friendly manufacturing practices (Starkey et al. 1998; Webster et al. 2003). Specifically, while recognizing the role of government regulations (Hoevenagel et al. 2007), work in this area also highlights the importance of firm characteristics and their associated effects on the transformation of business practices, investments in research and development, access to finance and relevant information as critical factors in the successful implementation of environmentally friendly manufacturing practices. On the whole these advances in literature have enriched our knowledge about environmental management at the firm level especially in relation to factors that lead to or impede environmental management among firms. Nonetheless, given that a majority of this research draws on the developed-country context and its major attention has been on large firms, there is still room for contribution especially in relation to how SMEs in developing countries adopt environmentally friendly manufacturing practices. This knowledge is critical and has the potential to enrich current advances in literature given that adoption of environmentally friendly manufacturing practices in the context of developing countries takes place within legacies of weak legal and regulatory frameworks, lack of resources and information and lack of demanding and sophisticated customers.

We now draw on extant literature to explain the likely relationships between an organization's characteristics of interest and adoption of environmentally friendly manufacturing practices and empirically establish how these characteristics may explain the environmental management of SMEs in a developing-country context.

2.1 The Role of Firm Age

Researchers have observed that with time, firms discover what they are good at and learn to be more efficient (Stewart and Amit 2003). They specialize and find ways to standardize, coordinate and speed up their production processes, reduce costs and improve quality. But it is also argued in some circles that older firms are prone to inertia, and are often constrained by the bureaucracy. Mature firms may therefore fail to make rapid adjustments to changing circumstances and are likely to lose out in performance stakes to younger and more responsive firms (Calantone et al. 2002; Duran and Regis 2001; Stewart and Amit 2003). In comparing young and old firms it is common for researchers to observe that young firms undertake radical green innovations, exploit technological or commercial opportunities more often and challenge the business models of existing firms more easily than older firms (Majumdar 1997). We speculate that similar logic may hold in the case of adopting environmentally friendly manufacturing practices among SMEs in a developing-country context.

2.2 The Role of Owner's Education

In addition to the age of a firm, personal demographic factors such as the age and education levels of the owner have a considerable impact on entrepreneurial intentions and/or decisions that can either hinder or promote the adoption of environmentally friendly manufacturing practices (Ki-hoon 2009; Schaper 2002; Walker et al. 2008). In particular, this is more likely among SMEs where ownership and management are concentrated in the same hands (Perez-Sanchez et al. 2003). In these firms, the personal preference of the owner will dominate any decision about investments, allocation of funds and the development of business strategies (Hoevenagel et al. 2007). Education is presumably related to knowledge and skills, motivation, self-confidence, problem solving abilities, commitment and discipline. As such, a higher education level of the owner is expected to increase his ability to cope with problems and seize opportunities that are important for the adoption of environmentally friendly manufacturing practices of a SME (Schaper 2002). Highly qualified managers are more likely to be innovative; they appear more likely to adopt strategies, introduce new higher quality products and improving the quality of existing products, while less qualified managers are less likely to be engaged in increasing the efficiency of the production of existing products and processes (Bosworth et al. 2000; Schaper 2002). All these are important aspects in the advancement of environmental management within a company. We also anticipate that similar logic may hold in the case of adoption of environmentally friendly manufacturing practices among SMEs in a developing-country context.

2.3 The Role of Owner's Age

Similar to an owner's education, his age may be linked positively to innovation (Amran 2011; Brockmann and Simmonds 1997). Thus, if we view adoption of environmentally friendly manufacturing practices as a form of innovation, we expect that age will negatively affect innovation. This is because mature people within innovation literature are noted to be more risk averse, while younger people change faster and are more inclined to change. Indirectly, these trends lessen chances of adoption of environmentally friendly manufacturing practices within a firm. Besides, younger individuals tend to be more environmentally concerned than older persons (Schaper 2002). Moreover, there is a clear relationship between environmental concerns and age with concerns rising in individuals till the 30–39 years age bracket and declining thereafter (Australian Bureau of Statistics 1998). Despite the alternative view that highly experienced managers are more likely to develop better production processes leading to adoption of environmentally friendly manufacturing practices, in the developing-country context we speculate that this relationship will potentially be negative.

3 Methodology

3.1 Research Design and Study Setting

This study explores environmentally friendly practices by SMEs in the manufacturing sector in the unique context of Africa. In addition, it establishes firm characteristics associated with those practices. The study is based on a survey of Ugandan

SMEs drawn from Mukono district during December 2013 and January 2014. Mukono district is located 21 km from capital Kampala. Mukono's proximity to the main economic hub of the country has made the district attractive for SMEs. Locating in Mukono district potentially implies easy access to the country's finest business development services including financial support that are available in the capital, yet at the same time avoiding the high location costs of the city center. In addition, Mukono's location near Lake Victoria and proximity to the Kenyan border, which represents a major transit for road imports and exports of the country, gives the district relatively easier access to the outside world. As a result, Mukono is now one of the fastest growing business hubs for SMEs in Uganda. Mukono district is therefore an interesting location to study in the context of Uganda. While we acknowledge that the SMEs in Mukono may have some unique contextual qualities as compared to SMEs located elsewhere in Uganda, we believe that given the highly dynamic environment in Mukono, SMEs in this district offer valuable insights in relation to the current research effort.

3.2 Study Sample

The sample was drawn from a number of manufacturing sectors including carpentry, steelworks, handcrafts and agro-processing. The focus was on firms located in Mukono town and in the Seeta industrial area. In this setting, and as also observed elsewhere in developing countries, local managers are afraid of losing their trade secrets and are skeptical of answering questionnaires for fear of leaking information to their competitors as well as fear of the information reaching the authorities. In such an environment, it was not possible to follow the procedure of random sampling. Instead a convenience sample was sought. Drawing on a convenience sample in developing countries is a common phenomenon given the lack of well-established databases of firms, the reluctance of executives to answer academic questionnaires and lack of accurate information of the firm's location (Aulakh et al. 2000). In line with the common practice, we worked with the local leadership of the two industrial areas and obtained a convenience sample of 122 SMEs. In terms of distribution across the two areas, 42 firms in the study were in the Seeta industrial park and 80 firms came from the Mukono town industrial park. Table 1 gives the characteristics of the study sample.

3.3 Data Collection

The data was collected using questionnaires. The second author personally delivered and picked the questionnaires from the respondents. She explained the study and clarified the necessary points as the respondents received the questionnaires. Given that the topic required a well established understanding of the firm's

Table 1 Sample characteristics

Variable	Category	Frequency	Percentage
Number of employees	1–5	72	59
	>5	50	41
Age of the firm	0–5	99	81
	6–10	23	19
Owner's education level	O-Level and below	32	26
	A-Level	35	29
	Diploma	24	20
	Degree	31	25
Owner's age bracket	18–34	68	56
	35–54	44	36
	>55	10	8

N = 122. The sample includes firms from both Seeta and Mukono town industrial areas

activities, we ensured that the questionnaire was received and answered by persons who were knowledgeable about the activities of the firm. Specifically the respondents were mainly owners and managers of SME establishments.

3.4 *Validity of the Study Responses*

In employing a survey data collection methodology, several issues arise that may compromise the validity of the study responses. Specifically, when a survey design is used, the findings of the study may be undermined by the problems of a selection bias, a non-response bias and common methods variance among others. When it comes to a non-response bias, it was not possible to conduct a post hoc examination given that there was no publicly available data to do a comparison test of the firms that did not participate in the study and those that did. Moreover, the questionnaires were collected in person therefore it was not possible to statistically ascertain the non-response basis by comparing early and late respondents. In order to reduce the occurrence of an informant selection bias, the questionnaires were personally administered by the second author. She ensured that the questionnaires were received and answered by the person who had necessary knowledge regarding the company.

In addition, the goal of the study rests on the validity of the behavioral construct of adoption of environmentally friendly manufacturing practices. A factor analysis with a varimax rotation procedure was conducted to establish the validity of the measures. Two dimensions were revealed that were labeled: recycling and nature of waste disposal. For each of these dimensions a reasonable loading of the variables retained is observed (see Table 2). In the case of the other study variables, the questions required objective data namely the age of the firm, the owner's level of education and owner's age. A detailed discussion of the variable measures is given later.

Table 2 Factor analysis with varimax rotation

Items/measures for adoption of environmentally friendly manufacturing practices	Factors F1	F2
We collect our waste to be taken by the municipal council		0.66
We have clearly labeled separate containers for hazardous waste		0.87
We have containers for different waste separate with clear labels and signs		0.82
Our products have recycled materials	0.73	
Scraps are reused or sold for reuse	0.87	
We collect back used material and use it in our production process	0.88	

Note: The underlying dimensions identified are two factors where F1 is recycling and F2 is waste disposal. The value for items that did load highly for each of the two factors is excluded. These results are based on factor analysis with varimax rotation procedure conducted in SPSS version 17.0

3.5 Study Measures

The dependent variable was environmental friendly practices adopted by the SMEs in the manufacturing sector. The study emphasized two dimensions: recycling and proper waste disposal. Highlighting these two dimensions helps one distinguish between adoption with immediate short-term financial value to the firm and that which is long-term and its value is primarily non-financial.

Recycling was measured using a five point likert scale consisting of three items: (a) the extent to which the firm used its scrap and damaged products with other raw materials in the production process, (b) the extent to which the firm's final products were made of reused materials, and (c) the extent to which a firm's policy was to collect back used products from its customers for use. These items were adapted from Eltayeb and Zailani (2009). The Crobanch Alpha coefficient for this scale was 0.77. The principle component for this scale revealed one factor with Eigen value of 2.09. The factor accounted for about 70 per cent of the variance. Individual factor loadings for the item were between 0.73 and 0.88 (see Table 2).

Waste disposal measures were adapted from Davis (2013). These consisted of three items measured on a five point likert measure. The items that made up this scale are the extent to which the firm (a) collects the company's waste to be taken by the municipal council, (b) clearly labels separate containers for hazardous waste, and (c) has acquired containers for waste separation with clear labels and signs. The Cronbach Alpha coefficient for this scale was 0.70. The factor accounted for 72 % of the total variance with factor loadings for individual items between 0.66 and 0.86. The independent variables in the study included the following firm characteristics: age of the firm, academic qualifications and age of the owner.

Table 3 Relationship between recycling, waste disposal, age of SME as well as owner's age and education level

Characteristic		Recycling		Waste disposal	
		Mean	Standard deviation	Mean	Standard deviation
Age of the SME	0–5	3.75**	1.07	2.20	0.78
	6–10	2.89	1.32	2.00	0.69
Age of the owner	18–34	4.03***	0.76	2.20	0.78
	35–54	3.09	1.36	2.21	0.76
	55–64	2.75	1.17	2.16	0.77
Education level of the owner	O-level	3.89	1.050	2.04	0.75
	A-level	3.65	1.136	2.23	0.92
	Diploma	3.52	1.249	2.13	0.74
	Degree	3.23	1.208	2.21	0.77

Note: **F significant at P < 01, ***F significant at P < 001

4 Results

The results of the study are based on means and standard deviations as well as an analysis of variance (ANOVA). The analysis was conducted using SPSS version 17.0. The cross-sectional nature of the data does not allow us to conclude on causality; as such the results are descriptive and only reveal associations. Table 3 shows the observed differences among SMEs' level of adoption of environmentally friendly manufacturing practices along the dimensions of the SME's age and the owner's education and age.

4.1 *Recycling, Age of the SME, Owner's Age and Education*

The results in this case reveal that recycling, which is a dimension of adoption of environmentally friendly manufacturing practices emphasizing immediate short-term financial value to the firm, may potentially be predicted by the age of the owner and the years of existence of the SME (see Table 3). With respect to the age of the owner, the data reveals that recycling activities among SMEs on average were higher among SMEs that were younger than those that were older. A similar trend is also observed for the relationship between recycling and age of the owner. On average, the level of recycling was noted to decrease with an increase in the owner's age. Drawing on ANOVA, both these observations are found to be significant at $P < 0.01$ and $P < 0.00$ respectively. While we observe a clear pattern of decline in the average level of adoption by the education level of the owner, this association is not found to be statistically significant. This means that this observation in the data may be out of chance. We were also interested in establishing which of these factors—age of the owner and years of existence—was potentially more likely to predict the level of recycling among SMEs in our study context.

Table 4 Relationship between recycling and organizational characteristics

Variable	Type III sum of squares	df	Mean square	F	Sign
Corrected model	2.19	5	5.84	5.128	001
Intercept	684.74	1	684.74	600.53	001
Age of SME	3.64	1	3.64	3.20	0.08
Owner's education	0.51	3	0.168	0.148	0.931
Owner's age	14.12	1	14.19	12.46	0.01
Error	107.00	94	1.138		
Total	1425.00	100			
Corrected total	136.19	99			

Note: (a) Computed using alpha = 0.05, (b) R square = 0.214 (adjusted square = 0.173)

Based on ANOVA, the results indicate that it is owner's age that is more likely to be associated with the level of recycling in our study context (for example, age of the SME estimate is $F(1, 107) = 3.20$, P = 0.008, while the age of owner is $F(1, 107) = 12.46$, P = 0.001). These results are presented in Table 4.

4.2 *Waste Disposal, Age of the SME, Owner's Age and Education*

While we found a strong association between recycling and the three factors of our interest in the study, similar findings were not established for waste disposal practices among SMEs in our context. None of the three SME characteristics (age of the owner, age of the firm and owner's education) was found to have a clear pattern of variation with the mean level of waste disposal management. It is therefore not surprising that the test of association for all the variables was found to be not significant. We discuss this finding in the next section and provide some insights into this observation.

5 Discussion, Conclusion, Limitations and Further Research

Based on the findings of the study it appears that SMEs emphasize immediate financial value when deciding whether or not to adopt environmentally friendly manufacturing practices. For instance, in the study none of the firm characteristics evaluated was found to be related to proper waste disposal practices—a dimension of environmental management which does not give obvious and immediate financial benefits. On the other hand, the age of the firm and the age of the management (captured by the owner's age) were found to be closely related to the adoption of environmentally friendly manufacturing practices. This finding is not surprising given that the legacies of weak regulatory frameworks will not encourage the already resource poor SMEs to incur extra costs on particular forms of environmental management that they consider not to be core to adding value to the organization. This finding also supports the view that given SMEs' resource constraints, there is need for governments in developing countries to address the regulatory machinery for SMEs to make necessary investments to safeguard the environment (Herren and Hadley 2010; Hoevenagel et al. 2007).

In our study, we offer an even richer guideline for governmental action. For instance, we find that there is already inherent motivation for SMEs to adopt certain environmentally friendly manufacturing practices with immediate financial value to the organization. We also identify the profiles of these SMEs. With this finding, governments in developing countries can save resources by specifically targeting those areas where SMEs are reluctant to adopt proper environmental management. We also highlight some of these areas in the study to guide the government's decisions (for example, this is in the area of waste disposal management).

Also interesting in the study is that the relatively young SMEs and those whose management is young were found to appreciate the value of recycling. This finding is consistent with earlier discussions in literature that talk of the negative relationship between age of the firm as well as the age structure of the management and adoption of environmental management (Labonne 2006; Simpson et al. 2004). While we recognize that there is a reverse argument in literature, we establish the former argument in this research as most likely to hold for SMEs in a developing-country context (Gadenne et al. 2009). It was, however, surprising to find that the relationship between educational differences among owners was not correlated with adoption of even those practices that would bring immediate financial value to an organization such as recycling. This is specifically surprising because one would expect educational differences to be associated with the level of awareness about environmentally friendly manufacturing practices and appreciation for the need to be environmentally friendly. This anomaly in our findings represents an interesting area for future research.

However, in interpreting the findings of this study scholars need to be aware of the following limitations. The study is only descriptive and does not provide causal explanations. It is based on a non-random sample and therefore it is not possible to make appropriate inferences about the population. Moreover, it is based only on a few organization characteristics and therefore addresses a limited area of the phenomenon. Notwithstanding these limitations, the study adds interesting insights to the literature on adoption of environmentally friendly manufacturing practices with potential to guide future research aimed at giving deeper insights into the current topic in the context of developing countries. In addition, the study offers some interesting guidelines for policymakers and firm managers to understand the nature of SMEs in the context of a developing country that are more likely to adopt to environmentally friendly manufacturing practices.

In conclusion, while this paper is a theoretical in nature, it contributes to our understanding on how certain key characteristics of an SME are associated with whether or not it will embrace environmentally friendly manufacturing practices in the context of a developing country. By distinguishing between manufacturing practices with immediate financial value to a firm and those that are not, the study gives a clear picture on how governments and firm managers within the legacies of weak legal and regulatory frameworks and firms with limited resources can effectively target their efforts at enhancing adoption of environmentally friendly manufacturing practices at the firm level. In line with the focus of the study therefore rather than test or extend the existing theory, we offer valuable insights which will help guide business practices and public policy.

References

Amran AN (2011) The effect of owner's gender and age to firm performance: a review on Malaysian public listed family businesses. J Global Bus Econ 2(1):104–116

Aulakh PS, Kotabe M, Teegen H (2000) Export strategies and performance of firms from emerging economies: evidence from Brazil, Chile, and Mexico. Acad Manag J 43(3):342–361

Australian Bureau of Statistics (1998) Environmental issues: people's views and practices. Australian Bureau of Statistics, Canberra

Avram OD, Kuhne S (2008) Implementing responsible business behavior from a strategic management perspective: developing a framework for Australian SMEs'. J Bus Ethics 82 (2):463–475

Berry AM, Randinelli AD (1998) Proactive corporate environmental management: a new industrial revolution. Acad Manag Exec 12(2):38–50

Biondi V, Frey M, Iraldo F (1997) Environmental management system implementation by SMEs: EU experiences and perspectives. Master's Thesis, Department of Environmental Management, North Carolina

Bosworth D, Davies R, Wilson R (2000) Managerial qualifications and organizational performance: an analysis of ESS 1999. Institute of Employment Research, University of Warwick, Coventry

Brammer S, Hoejmose S, Marchant K (2012) Environmental management in SMEs in the UK: practices, pressures and perceived benefits. Bus Strat Environ 21:423–434

Brockmann EN, Simmonds GO (1997) Strategic decision making: the influence of CEO experience and use of tacit knowledge. J Manag Issues 9(4):454–468

Calantone JR, Tamer CS, Yushan Z (2002) Learning orientation, firm innovation capability, and firm performance. Ind Market Manag 31(6):515–524

Davis G (2013) County (green) jobs in Queensland's waste and recycling sector. Waste Manag Res 31(9):902–909

Duran R, Regis C (2001) Age, order of entry, strategic orientation, and organizational performance. J Bus Ventur 16(5):471–494

Eltayeb TK, Zailani S (2009) Going green through green supply chain initiatives towards environmental sustainability. Oper Supply Chain Manag 2(2):93–110

Gadenne LD, Kennedy J, McKeiver C (2009) An emperical study of environmental awareness and practices in SMEs. J Bus Ethics 85(1):45–63

Global Environmental Management Institute (2004) Clear advantage. Global Environmental Management Institute, Washington, DC

Gutowski T, Murphy C, Allen D, Bauer D, Bras B, Piwonka T (2005) Environmentally benign manufacturing: observations from Japan, Europe and the United States. J Clean Prod 13 (2005):1–17

Hamdouch A, Depret MH (2012) Green entrepreneurship networks and clusters: when the local requires the global. RSA Global Conference, Beijing

Herren A, Hadley J (2010) Barriers to environmental sustainability facing small businesses in Durham, NC. Nicholas School of the Environment of Duke University, Durham, NC

Hoevenagel R, Brummelkamp G, Peytcheva AR (2007) Promoting environmental technologies in SMEs: barriers and measures. Institute for Prospective Technological Studies, Seville, Spain

Ki-Hoon L (2009) Why and how to adopt green management into business organizations? The case study of Korean SMEs in manufacturing industry. Manag Decis 47(7):1101–1121

Labonne J (2006) A comparative analysis of the environmental management, performance and innovation of SMEs and large firms. European Commission, Brussels

Lefebvre E, Lefebvre AL, Talbot S (2003) Determinants and impacts of environmental performance in SMEs. R D Manag 33(3):263–283

Majumdar SK (1997) The impact of size and age on firm-level performance: some evidence from India. Rev Ind Organ 12(2):231–241

Oba CV, Fobio I (2012) Board characteristics and the quality of environmental reporting in Nigeria. J Account Manag 2(2):33–48

Okello-Obura C, Minishi-Majanja KM, Cloete L, Ikoja-Odongo JR (2007) Assessment of business information access problems in Uganda. Can J Libr Inf Pract Res 2(2):1–34

Perez-Sanchez D, Barton JR, Bower D (2003) Implementing environmental management in SMEs. Corp Soc Responsib Environ Manag 10(2):67–77

Revell A, Stokes D, Chen H (2008) Small businesses and the environment: turning over a new leaf. Queen's University, Belfast

Schaper W (2002) Small firms and environmental management: predictors of green purchasing in western Australian pharmacies. Int Small Bus J 20(3):235–251

Seidel M, Seidel R, Tedford D, Cross R, Wait L (2008) A systems modeling approach to support environmentally sustainable business development in manufacturing SMEs. World Acad Sci Eng Tech 2(12):1305–1313

Seidel M, Seidel R, Tedford D, Cross R, Wait L, Hämmerle E (2009) Overcoming barriers to implementing environmentally benigh manufacturing practices: strategic tools for SMEs. Environ Qual Manag 18(3):37–55

Simpson M, Taylor N, Barker K (2004) Environmental responsibility in SMEs: does it deliver competitive advantage? Bus Strat Environ 13(3):156–171

Starkey R, Welford R, Young W, Brophy M, Rikhardsson P, Johnson C (1998) Environmental management tools for SMEs: a hand book. European Environment Agency, Copenhagen

Stewart T, Amit R (2003) Learning about failure: bankruptcy, firm age, and the resource-based view. Organ Sci 14(5):497–509

Walker B, Rednond J, Sheridan L, Wang C, Goeft U (2008) Small and medium enterprises and the environment: barriers, drivers, innovation and best practice. Edith Cowan University, Small and Medium Enterprise Research Centre, Perth

Webster B, Walker E, Schaper M (2003) Environmental management in small firms: is there a gender gap? Small Enterprise Res 11(2):63–70

Weesasiri SR, Zhang Z (2012) Impact of attitudes and awareness on environmental management practices (EMPs). J Asia Pac Bus Innov Tech Manag 2(2012):001–012

Williamson D, Lynch-Wood G, Ramsay J (2006) Drivers of environmental behaviour in manufacturing SMEs and the implications for CSR. J Bus Ethics 67(3):317–330

Are East African Countries Sustainable? Comparative Analysis of Two Composite Indicators

Miroslav Syrovátka and Jaromír Harmáček

Abstract This paper assesses the extent to which the development of five East African countries is sustainable. It starts with an overview of the concepts and measurement of sustainability, focusing on composite indicators. The main part of the paper presents an analysis of two composite indicators (Adjusted Net Savings and Ecological Footprint) as applied to five East African countries (Burundi, Kenya, Rwanda, Tanzania and Uganda). For each of the two indicators, we provide a short overview of the concept, a critique and an analysis of the results and policy implications. The results show that only one of the countries is unsustainable in Adjusted Net Savings, while the Ecological Footprint shows either all five countries as sustainable or all five unsustainable, depending on the interpretation of bio-capacity. Since the indicators lead to different conclusions on countries' sustainability, we analyse the implications and discuss to what extent these indicators can be used for assessing country sustainability.

Keywords East Africa • Sustainability • Composite indicators • Adjusted net savings • Ecological footprint

1 Introduction

A major challenge of this century is how to make development sustainable. The term 'sustainable development' has been in use since the famous book *Our Common Future* (WCED 1987) that defined it as a 'development that meets the needs of the present without compromising the ability of future generations to meet their own needs.' Since 1987 we have seen development in most parts of the world, measured by various indicators of socioeconomic development. An average person in the world today has more income, higher education levels and is in better health than ever before. Also, though we do not have historical estimates of subjective

M. Syrovátka (✉) • J. Harmáček
Department of Development Studies, Palacky University, 17. listopadu 12, 771 46 Olomouc, Czech Republic
e-mail: miroslav.syrovatka@upol.cz; jaromir.harmacek@upol.cz

A. Heshmati (ed.), *Economic Integration, Currency Union, and Sustainable and Inclusive Growth in East Africa*, Advances in African Economic, Social and Political Development, DOI 10.1007/978-3-319-30432-8_9

well-being except for the most recent period, it seems that it, after all, increases with income (see Stevenson and Wolfers 2013). Not every person and country is better off than a quarter century ago in every aspect of life, but on average most people and countries have progressed. What is less clear, however, is whether this progress has been sustainable, that is, whether it has not compromised the development opportunities of future generations.

The aim of this paper is to analyse the sustainability of five East African countries (Burundi, Kenya, Rwanda, Tanzania and Uganda) using two composite indicators. The Classification of Composite Indicators of Sustainability section summarizes the concepts of sustainability and the discussion over its measurement by composite indicators. Adjusted Net Savings and Ecological Footprint sections analyse the results and policy implications for East African countries in the Adjusted Net Savings (ANS) and the Ecological Footprint (EF) indicators. We analyse whether the indicators provide a similar assessment of sustainability in East Africa and taking into account the limitations of these indicators what the policy implications are.

2 The Classification of Composite Indicators of Sustainability

Measuring sustainability is an important part within the concept of sustainable development. In 1992, Agenda 21 adopted in the United Nations Conference on Environment and Development, expressed the need to establish indicators in order to better monitor and foster sustainable development. Since the 1990s there has been an emerging interest in constructing indicators that measure certain aspects within the concept of sustainable development.[1] These indicators may be divided into three groups based on the degree of aggregation.

On the lowest hierarchical level are wide sets of less aggregated indicators. A typical example is the structure of sustainable development indicators of the European Union (EU). This is a set of 130 indicators on various aspects of sustainable development. These indicators may provide a complex assessment; however, their disadvantage lies in a complicated presentation and difficult interpretation of the overall trend from a set of separate indicators. To make the interpretation easier, we may make use of composite indicators that combine several aspects into a single number (for a general guide for constructing composite indicators see OECD 2008). If sets of separate indicators are at one end of the spectrum, composite indicators of sustainable development are on the other end. These indicators embrace aspects of both sustainability and development and

[1] For an overview of the measurement of socioeconomic progress and sustainability see Stiglitz et al. (2009). Bandura (2008) provides an annotated list of composite socioeconomic indicators (though it is not up-to-date given the fast development of new indicators).

combine them in a single number. These indicators are represented by both composite indices (see two types of sustainable development indices by Barrera-Roldán and Saldívar-Valdés 2002 and by Mederly et al. 2002) and monetary indicators such as the Index of Sustainable Economic Welfare (Daly and Cobb 1989) which is today often called the Genuine Progress Indicator (see, for example, Talberth et al. 2006). Though composite indicators are constructed to make it easier to interpret certain phenomena and trends, here the drive towards a simplified interpretation is contra-productive and leads to less, rather than more, interpretation value. Using the analogy by Stiglitz et al. (2009), when driving a car, a single meter measuring speed and the level of gasoline would be of no use to the driver who needs two separate meters for measuring two distinct phenomena.

Somewhere between the two ends of the spectra are composite indicators of sustainability and composite indicators of development. These are still highly aggregated indicators, but since they aggregate only within each of the concepts (sustainability and development), they allow for reasonable interpretations. Sustainability indicators are much younger than widely defined indicators of development and quality of life. Indeed, such an eminent economist as Robert Solow argued at the beginning of the 1990s that 'sustainability is an essentially vague concept, and it would be wrong to think of it as being precise, or even capable of being made precise ... It is intrinsically inexact. It is not something that can be measured out in coffee spoons. It is not something that you could be numerically accurate about' (Solow 1991: 180, 187). Yet, it was in the 1990s when the first composite indicators of sustainability were developed. These indicators can be categorized into three groups that represent three approaches to measuring sustainability: indicators based on a monetary valuation, bio-physical indicators and composite indices.

Monetary indicators of sustainability reflect the economic approach to sustainability. This approach understands sustainability as non-declining welfare (or utility) generated by capital. Therefore, if the stock of capital does not decrease over time, welfare need not decline. Capital is measured in monetary units which is both a strength and a weakness of this method. There is a longstanding theory in economics to express the value of goods and services in prices that are established in markets. Yet, not all types of assets relevant for sustainability are traded in markets to allow us to use market prices. And even if prices are established in markets, it is not necessarily the case that market prices are relevant for sustainability as they may not properly reflect scarcity. We can distinguish several types of human-made capital (physical capital, human capital and possibly also social capital) and natural capital.[2]

Two approaches to sustainability have evolved with respect to this. In a weak sustainability paradigm, various types of capital can substitute one another. The general policy of sustainability is that of maintaining the non-declining value of

[2] Be aware of the difference between human capital and human-made capital. Human-made capital is any capital created by humans, while human capital is one of the sub-groups of that capital.

total capital. As a practical example, if we use certain non-renewable natural resources (natural capital) and invest the resource rents in education (human capital), this is considered sustainable since we replace one type of capital with another. The Adjusted Net Savings developed under the auspices of the World Bank (1997) is a representative indicator of this category. In a strong sustainability approach, certain types of capital cannot be replaced by other types of capital. This irreplaceable capital is usually understood as natural capital generally or as certain types of natural capital. Strong sustainability implicitly rejects the monetary assessment of sustainability based on unlimited substitution. Instead, indicators of strong sustainability are based on the concept of bio-physical carrying capacity of the natural environment to function as a source and sink.

The conceptual foundations for *bio-physical indicators* can be found in Hardin (1976), Ehrlich and Holdren (1972) and Catton (1980). Later operationalizations of the concept of bio-physical carrying capacity include human appropriation of net primary production (Krausmann et al. 2013; Vitousek et al. 1986), material and energy flows accounting (Eurostat 2012) and the Ecological Footprint (EF). The defining characteristic of bio-physical indicators is that they directly take into account environmental resources as the only aspect of sustainability. EF is one of the most developed bio-physical indicators (and probably the most popular indicator in the environment-sustainability area). Also, unlike other bio-physical indicators that measure human impact on the environment, EF allows for sustainability assessment. This is because beyond measuring human impact the EF concept has been designed also to measure carrying capacity that provides a threshold with which the human impact can be compared. In this analysis we therefore use EF, as it is the only well-known bio-physical indicator of sustainability.

The third approach towards measuring sustainability is served by *composite indices*.[3] Developers of indicators often resort to indices in case they do not see another more 'natural' way of measuring the given concept. The index is an indicator composed of several parts that are normalized, weighted and aggregated. Since there are no strict rules for selection of sub-indicators, normalization, weighting and aggregation, the choices made by the creators of the index are crucial. This gives rise to a critique that indices are not grounded in theory, and therefore, arbitrary and worthless.[4] Though we acknowledge the difficulty of constructing a methodologically sound composite index, we think that the approach of considering all composite indices a-priori worthless is too strict. Measuring market activity by a composite index may not bring any additional value when reasonable theoretically grounded monetary indicators are available. However, when such a solid theoretical background is missing, a composite index may be

[3] The terminology for classification of various summary indicators is not unified. We use 'composite indicator' as a broad term encompassing all types of highly aggregate indicators and reserve the term 'composite index' (plural 'indices') for only those indicators measured on a dimensionless scale.

[4] See, for example, Ravallion (2012) for a critique of composite indices of development.

useful. Yet, the only well-known composite index of sustainability—the Environmental Sustainability Index—is no longer updated. It was created in 2000 but its development was terminated after few years (see Esty et al. 2005 for the last version) and it was replaced by the Environmental Performance Index (see Hsu et al. 2016 for the latest version). Neither of these indices is used in our analysis. The former is no longer used as a relevant measure of sustainability since it has been replaced and has not been updated (most of the underlying data is from the 1990s), while the latter deliberately does not attempt to measure environmental sustainability; it measures environmental performance instead.[5] Moreover, neither of them provides a clear threshold for assessing sustainability.

In the following two sections, we analyse two composite indicators (Adjusted Net Savings and the Ecological Footprint) as applied to five East African countries (EA5). For each of the two indicators, we provide an overview of the concept and a critique and analyse the results and policy implications. As for the results, we contrast East African countries with other country groups and where relevant we emphasize the differences between the five countries.[6]

3 Adjusted Net Savings

3.1 Concept

The Adjusted Net Savings (ANS) indicator is based on a capital approach to sustainability.[7] What we want to sustain here is consumption (or, more generally, utility or welfare derived from consumption) for which we need production. The core concept behind ANS is that of capital. For economists, capital has always been essential since its accumulation allows non-declining consumption and utility in the future. Various types of capital are used in production (not only physical capital traditionally understood as *the* capital) and ANS is based on the assumption that these types of capital can substitute each other. Thus, we are not interested in any individual type of capital, but only in capital in an aggregate sense. Should various types of capital be comparable, they need to be expressed in same units, which in the case of ANS are monetary units.

ANS starts with gross savings and adjusts these by items that affect the stocks of different types of capital (see Bolt et al. 2002; World Bank 2011). For calculating gross national savings (GNS) we deduct public and private consumption from gross

[5] It can be argued that neither ANS nor EF have 'sustainability' in their names. However, they are commonly interpreted as indicators of sustainability.

[6] Data for both indicators are for 2011. This was the most up-to-date data for EF at the time of writing (mid-2015). While ANS data are available for more recent years, we used the same year for consistency.

[7] ANS is also called genuine savings, a term coined by Hamilton (1994).

national income (GNI) and add net current transfers. Four types of adjustments then follow. First, from GNS we deduct depreciation of physical capital (that is, replacement value of capital used in the production process) to obtain an indicator of net national savings (NNS). This follows the simple logic that if we save more than the capital depreciates, we can always replace the decrease in capital stock by investing the saved resources. If physical capital is the only capital necessary for sustainability, then this is the rule of sustainability. Since both human capital and natural capital are also important for sustainability, this necessitates further adjustments. Second, we add education expenditures. In standard national accounting, these expenditures are counted as consumption, though they in fact constitute investments in human capital. Since the weak sustainability paradigm assumes unlimited substitution between various types of capital, an increase in human capital stock can offset a decrease of other types of capital. Third, we subtract rent from the depletion of several types of natural capital that reflect a decrease in the value of their stocks. This includes mineral and energy resources as non-renewables and forest resources as renewables.[8] Finally, we subtract pollution damages, that is, damages caused by carbon dioxide (CO_2) and particulate matter (specifically, PM10).

The calculation can be summarized by a formula divided into four parts as described earlier:

$$\mathrm{ANS} = (\mathrm{GNS} - \mathrm{CFC}) + (\mathrm{EE}) - (\mathrm{NFD} + \mathrm{ED} + \mathrm{MD}) - (\mathrm{CO_2} + \mathrm{PMD})$$

where GNS stands for gross national savings, CFC for consumption of fixed (physical) capital, EE for education expenditures, NFD for net forest depletion, ED for energy depletion, MD for mineral depletion, CO_2 for carbon dioxide damage and PMD for particulate matter damage.

Total ANS can be expressed in monetary terms and as a share of GNI. If ANS is not negative, we leave future generations with at least the same aggregate level of capital as we have, and so we are on a sustainable path. Should ANS be negative, we are unsustainable since the total capital stock is declining. Unlike EF that focuses on environmental resources, ANS focuses on production opportunities. It does not require a country to keep nature as it is, not even to keep natural capital in some aggregate sense. ANS requires us to keep whatever is necessary for future production, that is, the total capital stock.

[8] Mineral resources include tin, gold, lead, zinc, iron, copper, nickel, silver, bauxite and phosphate. Energy resources cover coal, crude oil and natural gas. Unit rent is calculated as the world price of extracted resource minus the average unit extraction costs (which include 'normal profits'). Total rent (for each resource) is calculated as unit rent times production volume. For forest resources, rent is calculated only from the extraction that exceeds natural growth of the wood matter.

3.2 Critique

ANS is an indicator of weak sustainability. The conceptual critique of ANS is not specifically about the computation itself but relates to the general assumption of weak sustainability. Some economists (mainly ecological economists) and many environmentalists argue that the substitution of human-made capital for natural capital is very limited since for production purposes these are rather complements than substitutes (see, for example, Costanza and Daly 1992). If this is the case in principle, then any indicator of weak sustainability misses the point, and we need to monitor various types of capitals independently.[9] We will not go much further in this discussion, except to note that reasonable middle ground acknowledges that substitution is more feasible for natural resources than for environmental services such as global climate regulation.

The two specific shortcomings of ANS relate to what is counted as capital and on the price of that what is counted (Stiglitz et al. 2009). As for human capital, the current education expenditure is only a proxy for an increase in human capital stock. Since it is calculated as the cost of providing education, it does not take into account the type of education, its quality and other factors that might be relevant for human capital accumulation. Human capital is also affected by health, which is not considered by ANS. According to one estimate, health capital may be larger than all other forms of capital combined (see Arrow et al. 2012).

Some types of natural capital depletion are not counted like the many types of mineral resources, fisheries, soil and most of the pollutants. Conceptually, these items may be included in ANS, but practically, this is difficult due to lack of internationally comparable data and the pricing issue. While there are market prices for natural resources, these may be understood as 'correct' prices for sustainability measurement only when certain conditions are met (for example, complete, competitive markets with no externalities). This is rarely the case in reality, and so the measured ANS differs from the true ANS computed from theoretically 'correct' prices.[10] Also, the fluctuations in prices of some mineral and energy resources affect ANS which limits the practical relevance of this indicator for some countries (Stiglitz et al. 2009). As for forest depletion, the calculation is based on market prices of net loss of wood. However, some indirect but long-term effects of deforestation such as soil erosion and increased risk of flooding are not captured by ANS.

[9] There are various approaches towards strong sustainability that require us to maintain one of these: (a) the value of each type of capital separately as none of them is mutually substitutable, (b) the value of natural capital as it is non-substitutable for any type of human-made capital (but various kinds of natural capital can substitute one another), and (c) physical stocks of certain types of natural capital (so-called critical natural capital) as these cannot be substituted for any other capital.

[10] There is also a link between prices and substitutability. Even if substitution is possible, the substitution is not necessarily in relation to prices used for calculation of ANS (see Neumayer 2013).

Only two types of pollution are included (CO_2 and particulate emissions) which underestimates the total pollution damages in ANS. The valuation is even more difficult since there are no market prices. For example, damages from CO_2 emissions are currently valued at US$ 20 per ton of carbon (1995 US$), but in the light of newer estimates this might significantly underestimate the true social cost of carbon. In that case, the measured ANS will be overestimated in countries with high emission intensity of production. The low social cost of carbon may send wrong signals to policymakers that carbon emissions are not *that* important for sustainability. There is one subtle, but important point related to emissions with transboundary impacts. The CO_2 damage in ANS is assigned to an emitting country, though in reality, the country's damage (decrease of country's capital) depends on global emissions and the country's characteristics, not on national emissions. This shows that country sustainability is not affected by domestic affairs only and points to the difficulty of defining sustainability on a country level.

3.3 Results and Policy Implications

Table 1 summarizes the results of ANS for EA5 countries and selected country groupings. When considering only standard national accounting, four countries show positive net national savings (GNS minus CFC), while it is negative for Burundi (−3.7 %; all percentage values for ANS and its components relate to GNI). When net national savings are adjusted for other factors affecting sustainability, a much higher unsustainable pattern is revealed in Burundi (−24.0 %), while the remaining four countries keep their ANS positive.

We analyse and discuss the results of EA5 countries in individual components and in ANS as a whole. We also provide a comparison with one regional aggregate (Sub-Saharan Africa) and three income groups. Since all EA5 countries were low income countries (LICs) according to the World Bank's classification, we compare them with an average for this group and for contrast we also use the other two income groups—middle income countries (MICs) and high income countries (HICs).[11]

The main message of ANS is that most of the world's countries are sustainable. In 2011 (The World Bank 2015a), there were 20 countries with negative ANS and 104 countries with positive ANS. On average, the world is on a sustainable path (ANS 10.0 %) as are each of the three income groups.[12] This is also true for our regional aggregates—EA5 (6.2 %) and Sub-Saharan Africa (2.7 %). Regional

[11] In 2015 Kenya has been reclassified as a lower middle income country as its GNI per capita for 2014 exceeded 1,045 US$.

[12] While low and high income countries show an average ANS of around 7–9 %, middle income countries tend to be more sustainable with an average ANS of 16 %. A dominant factor explaining this difference is GNS, that is, MICs save much more than both LICs and HICs. Irrespective of income, economies based on extraction of natural resources tend to be unsustainable.

Table 1 Adjusted Net Savings (% of GNI, 2011)

	GNS (+)	CFC (−)	EE (+)	ED (−)	MD (−)	NFD (−)	CO_2 (−)	PMD (−)	ANS
Burundi	9.8	13.5	5.8	0.0	0.9	23.6	0.1	1.5	−24.0
Kenya	10.9	6.5	5.9	0.0	0.1	3.2	0.3	1.0	5.6
Rwanda	20.9	11.1	4.1	0.0	0.2	5.7	0.1	1.0	6.8
Tanzania	19.6	7.7	4.8	0.4	3.9	0.0	0.2	0.5	11.6
Uganda	19.6	5.9	2.9	0.0	0.2	13.7	0.2	0.6	1.8
EA5	15.9	7.2	4.9	0.1	1.4	4.7	0.2	0.8	6.2
SSA	21.9	9.1	3.7	9.2	1.8	1.6	0.5	0.8	2.7
LICs	23.1	7.5	3.1	1.1	1.6	4.5	0.3	1.2	9.4
MICs	34.1	12.8	3.3	5.5	1.5	0.3	0.8	0.3	16.2
HICs	21.5	12.8	4.7	1.9	0.2	0.0	0.3	0.1	7.5
World	25.1	15.1	4.3	2.9	0.6	0.1	0.4	0.2	10.0

Source: The World Bank (2015a); authors' calculations
Note: GNS: gross national savings, CFC: consumption of fixed capital, EE: education expenditure, ED: energy depletion, MD: mineral depletion, NFD: net forest depletion, CO_2: carbon dioxide damage, PMD: particulate matter damage, EA5: five East African countries, SSA: Sub-Saharan Africa, LICs: low income countries, MICs: middle income countries, HICs: high income countries. Total values for country groups are GNI-weighted averages. Numbers may not add up due to rounding off

aggregates hide notable exceptions to the general pattern. While EA5 countries are on average sustainable, one country (Burundi) is largely unsustainable (−24.0 %). Given that a dominant factor behind Burundi's unsustainability is negative NFD, the easy policy recommendation for the country is to decrease the rate of deforestation (or to invest all the resource rents in physical and human capital), while the other four EA5 countries do not need to do anything substantial. However, accepting the general concept of ANS and taking into account that its calculation is not without problems, we may disaggregate total ANS into its individual components and provide a more nuanced picture.

The starting point of ANS is GNS. East African countries have their GNS (15.9 %) lower than any income group and also than that of Sub-Saharan Africa as a whole. This is mainly due to low GNS of around 10 % in Kenya and Burundi. This is an important point since GNS is the core of ANS with a higher magnitude than the other factors. Sustained economic development of a country is dependent on its ability to save a fraction of national income that can be used for investments. The government has a role in creating an environment that leads private agents to increase their savings by investing in infrastructure, making business easier and predictable, maintaining stable financial system etc. Yet, concrete policy advice has to be adjusted to the context of a particular economy. For example, Rwanda already has a favorable environment for business vis-à-vis the other four countries, so it is likely that a different policy focus will lead to increasing GNS in these countries.[13] Country specific conditions matter and the differences in countries' GNS are not

[13] While Rwanda is 46th on the World Bank's ease of doing business index (which is almost on par with South Africa and the Czech Republic), the other four countries are between the 131st and 152nd place out of 189 countries (The World Bank 2015b).

always easily attributable to standard explanatory factors. For example, Kenya has low GNS, yet it is not generally lagging behind other East African countries in many factors that are generally considered relevant for saving decisions.[14]

Four EA5 countries show a solid performance in public investment in education. East Africa's average expenditure (4.9 %) on education is significantly higher than both Sub-Saharan Africa's average (3.7 %) and LICs' average (3.1 %). While we acknowledge that allocating such share of the country's resources to education (with the exception of Uganda) is a significant political achievement vis-à-vis other SSA countries, it masks two aspects. First, while EA5 provide slightly higher percentages of GNI for education than HICs (4.7 %), poor countries have younger population structures that necessitate *higher* education expenditure (at least in a relative sense as a share of GNI). Second, the current education expenditure is only a proxy for an increase in human capital stock. Both data and anecdotal evidence show very poor quality of education in many developing countries, including in East African countries. With a gradual increase in enrollment rates in primary and secondary education in the last decades the more pressing need may be the quality of the education process itself.

East African countries show relatively low mineral depletion and only negligible energy depletion compared to other LICs. While some African countries have significant reserves of energy resources (mainly oil) that are being extracted, countries in the Eastern part of Africa generally lack these resources. Tanzania is the only exception to this pattern, mainly due to gold extraction and newly discovered natural gas deposits (estimated at over 50 trillion cubic feet), the first of which were discovered in 2010 in the Indian Ocean.[15]

Regarding the management of natural resources, there are two essential questions for governments of resource-extracting countries. First, whether they can capture a sizeable share of resource rents (profits) in corporate taxes and royalties, without discouraging existing and potential (foreign) investors. As the price of gold has risen by 5 % in real terms per annum since 1990s, royalties have become a lesser burden for most gold miners. Some African governments used this opportunity to increase the royalties (Gajigo et al. 2012). Tanzania updated its mining regulations in 2009–2010, including an increase in royalties on gold from 3 to 4 % and cutback of some tax concessions for mining. By renegotiating existing contracts with mining companies and amending the legislation, the governments may capture a higher share of the resource rents (see Muganyizi 2012 for policy recommendations

---

[14] Within the EA5, Kenya is not last either on the ease of doing business index (The World Bank 2015b) or in any of the six dimensions of Worldwide Governance Indicators (The World Bank 2015c). Also, Kenya is not last in Logistics Performance Index (The World Bank 2015a) that evaluates the quality of trade and transport related infrastructure (newest year available for all indices).

[15] While Tanzania is only the world 15th major gold producing country (with a 1.7 % share of global gold production as compared to China's 14.4 %) it is 3rd in the share of gold in the national economy (6 %) and number of employees (17,100) and first in the share of gold in national merchandise export (36 %). (All data are for 2012 and come from PricewaterhouseCoopers 2013.)

for Tanzania). This is generally easier for contracts relating to newly discovered deposits, such as for Tanzania's natural gas (see The Economist 2014).

The second question is how the revenues are actually used to benefit people now and in the future. While the strict logic of ANS sustainability requires *all* the resource rents to be invested, an argument can be made that in poor countries these resources should be shared between current and future generations. Part of the funds may even be distributed directly to people to limit rent-seeking or they can be earmarked for cash transfer programs (see Gupta et al. 2014 for an overview). The other part of the funds may be saved, possibly in foreign denominated assets (for example, via sovereign wealth funds) to limit the negative effects of domestic currency appreciation. These two questions are inherently linked by the issue of transparency. There are international initiatives that try to promote transparent and accountable management of natural resources. Probably the most well-known of them, the Extractive Industries Transparency Initiative (EITI), ensures that any payments made by oil, gas and mining companies to governments are disclosed in annual reports open to anybody, but primarily to the citizens of the country. The most resource-extracting country, Tanzania, complies with EITI standards, the other four are not involved in it. For diamonds, the Kimberley Process Certification Scheme (KPCS) should ensure that 'conflict diamonds' do not enter the rough diamond market. As of now, Tanzania is a participant in the scheme, while Kenya expressed commitment to the Kimberley Process but has not yet met its minimum requirements.

Forest depletion is a significant issue for many LICs. Due to very high forest depletion in Burundi (23.6 %) and Uganda (13.7 %) the average for EA5 countries (4.7 %) is three times higher than the average for SSA (1.6 %) and it contrasts with almost zero net depletion for the world as a whole (0.1 %). Over the period 1990–2012 the forest area in the two countries decreased by more than 40 %, from 11.3 to 6.6 % of the land area in Burundi and from 23.8 to 14.1 % of the land area in Uganda (The World Bank 2015a). Forest depletion is a consequence of population pressure, poverty and lack of effective regulations. Forests are converted to agricultural land (see, for example, Butamira and Bugala Islands in Uganda) and wood is used as a source of energy and timber.[16] Forest depletion is an illustrative case of a trade-off between current well-being and sustainability (that is, a potential for future well-being). If an area is deforested and the proceeds are not reinvested, current needs are satisfied at the expense of future generations.

All five countries show relatively small CO_2 emissions per capita.[17] However, as emission damages (as well as all other components of ANS) are measured in relation to GNI, the EA5 emission damages (0.2 %) are not so much different from those in HICs (0.3 %) that have significantly higher emissions per capita.

[16] See Obua et al. (2010) for more details on forests in Uganda.

[17] From the EA5, the lowest CO_2 emissions are for Burundi (0.02 tons) and the highest for Kenya (0.28 tons). Compare this with 0.7 tons for Sub-Saharan Africa, 9.4 tons for Germany and 17.1 tons for the US. All data relate to 2011 and exclude land use, land-use change and forestry (WRI 2015).

The recent history of industrial development in fast-growing middle income countries shows a trend of increasing per capita emissions. Changing that course in developing countries requires effective domestic regulations and transfer of technologies and funds from developed countries.

4 Ecological Footprint

4.1 Concept

EF is an indicator of human consumption of renewable resources. Though EF represents an original method of quantification of human use of natural resources, it builds on older concepts that relate human consumption of natural resources to the limited capacity of the natural environment. It builds on the concept of carrying capacity, but implicitly defines it as a maximum load that can be safely loaded on the environment (Catton 1980), taking into account population, consumption and technology (Ehrlich and Holdren 1972). The essential question of EF is that of human appropriation (Vitousek et al. 1986), that is, how much of the carrying capacity is appropriated by humans. Indeed, one of the originators of the concept proposed to measure the sustainability of society inhabiting certain areas with 'appropriated carrying capacity' (Rees 1992), later renamed the 'Ecological Footprint'. The methodology was first presented in the book *Our Ecological Footprint* (Wackernagel and Rees 1996) and has evolved over time.[18]

EF's aim is to quantify human use of natural capital and compare this use with natural capital that is produced on a given area (usually a country) for a given time (usually a year). Unlike ANS, this analysis is not carried out in monetary units but in bio-physical units—global hectare (gha)—a unit that represents a hectare of biologically productive area with world average productivity. More specifically, EF measures the area of biologically productive land and water that is needed for the production of renewable resources consumed by the human population and for the assimilation of waste generated by it. EF is a measure of human demand on natural sources and sinks. Should EF be used to assess sustainability, it needs to be compared with bio-capacity, that is, with the area of biologically productive land and water available for a given population. We supposedly assess whether a given population lives within the carrying capacity of its environment (EF is lower than bio-capacity) or not (EF exceeds bio-capacity). By dividing EF and bio-capacity by population, the results can be expressed in per capita terms.

[18] For an overview of methodology and results see Borucke et al. (2013). Updated information can be found on the website of the Global Footprint Network.

4.2 Critique

The EF concept looks intuitive and is appealing as a communication tool for showing human demands on nature. However, both the concept and methodology of EF have been challenged.[19] Regarding the methodological challenges, there are arguments that the calculated EF is both underestimated and overestimated for different reasons. Underestimation stems from the fact that EF does not cover all types of environmental degradation, but only those for which there is a substantiated method of conversion into a biologically productive area. For example, EF does not include depletion of all non-renewable resources since there is no biologically productive area that reproduces them in our timescale. It also does not include any type of pollution except CO_2 emissions ignoring, for example, emissions that deplete the ozone layer and all toxic waste.

Another point is that out of five types of biologically productive areas, the methodology does not allow either deficit or reserve to occur on three of them. For cropland, grazing land and built-up land, the global footprint equals global bio-capacity.[20] While the consumption footprint on cropland may differ from cropland bio-capacity on a country level, the production footprint always equals bio-capacity. Agricultural practices that will lead to worse soil quality and lower productivity in the future, will not be reflected in the current ecological balance (while this will be reflected in future in lower bio-capacity *and* a lower footprint, again, there will be no reserve or deficit). Globally, unsustainability can originate only on two types of areas—fishing ground and forest land. Forest land is the only area in the EF concept that can cover two types of human footprint—forest as a source of wood and as a sink for carbon (see details later). Assessing the sustainability of the human use of fish and wood, the methodology evaluates how current consumption affects the stock of fish and wood. Since this reflects basis sustainability principles and in the global level these footprints are lower than bio-capacity (Stiglitz et al. 2009), global unsustainability is driven by the carbon footprint whose methodology has been questioned.

Overestimation of EF results from a calculation method that produces a hypothetical scenario that does not take into account technological progress. More specifically, fossil fuel use is converted to a carbon footprint, that is, an area of forest needed to sequestrate the emitted CO_2. This arguably overestimates EF since it does not consider other options for conversion of fossil fuel use to a biologically productive area, options that might be expensive yet economical in terms of the

[19] From the older papers, see for example, van den Bergh and Verbruggen (1999). For a recent discussion on the methodology and policy relevance of the EF concept, see the critique by van den Bergh and Grazi (2013), response by Wackernagel (2014) and consequent response by van den Bergh and Grazi (2014).

[20] This is true on a global level, while for countries cropland and grazing land footprints may differ from respective bio-capacities. Footprint and bio-capacity on built-up land always equal even on a country level. For cropland, there can be a deficit on a country level, but not on a global level.

required biologically productive area.[21] Since the carbon footprint constitutes 55 % of global EF (and 60 % in HICs), the method of conversion has a substantial effect on the results. Also, this hypothetical scenario offers an interpretation that afforestation is a sustainable long-term scenario which it is not.[22]

The EF concept is generally considered to be an indicator of strong sustainability. As it measures how much biologically productive area do we use (and compares it with how much do we have), it apparently takes into account only natural resources. Yet, it implicitly takes into account human-made capital as well. When EF researchers convert consumption into a biologically productive area based on 'prevailing technologies and resource management' of a given year for which the EF is calculated (Borucke et al. 2013: 519), they implicitly consider a certain level of human-made capital that defines prevailing practices and technology. However, unlike indicators with explicit measurements of some types of human-made capital such as ANS, EF factors them in indirectly. The change in human-made capital is reflected in EF calculations only when the growth of human-made capital transforms into a new (higher) level of 'prevailing technologies and resource management'. Since what is prevailing is not defined, it allows authors to use forest sequestration as a basis for the carbon footprint.

Besides methodological problems, there is another issue concerning the spatial aspects of sustainability as measured by the EF concept: at what level should EF be applied and what it should be compared with. Given the unequal distribution of the world population and natural resources, it is questionable whether we should expect the Netherlands to live within its bio-capacity of 1.1 gha per capita (all values for footprints and bio-capacities below are in per capita terms). With a moderately high EF (4.5 gha), the Netherlands shows a negative ecological balance (a deficit of −3.3 gha), while Canada shows a very large positive balance (a reserve of 8.0 gha) even with very high EF (6.6 gha). Indeed, it would be difficult for Canada *not* to live within its large bio-capacity (14.6 gha). As van den Bergh and Verbruggen (1999) have argued, given the historical pattern of population density and allocation of environmental resources, it does not make sense to ask populations of small, densely inhabited areas (such as the Netherlands and London) to live within carrying capacities of their environments (country and city). For the authors, EF applied on a country level has an anti-trade bias. It does not allow for net import of bio-capacity above the 'consumed' domestic bio-capacity, even if it is imported from a country with an ecological reserve.[23] It is therefore questionable whether the EF concept has meaningful applications at a lower than global level.

---

[21] Think of using energy from solar panels based on non-productive land such as buildings (see Neumayer 2013).

[22] Trees first absorb carbon but later release it once they decay naturally or are burnt by humans. Taking account of the full carbon cycle, afforestation does not affect the carbon balance in the long term.

[23] If EF is compared with the average global bio-capacity (rather than national bio-capacity), this problem diminishes. Yet, countries with large bio-capacity per capita may claim their political sovereignty over the natural resources on their territories.

4.3 Results and Policy Implications

To assess East Africa's performance comprehensively, we show both interpretations of the EF concept. The results are summarized in Table 2. Assessing the East African countries solely in terms of their EF per capita, their demands on human nature span from 0.8 gha in Burundi and Rwanda to 1.2 gha in Tanzania and Uganda, which is in line with other African countries as well as other LICs. While an average person in the world needs 2.7 gha of biologically productive area for his consumption, the average East African person needs much less (1.1 gha). There is a general pattern for EF to increase with economic levels. EF of HICs (5.1 gha) is five times that of LICs (1.0 gha) and the Pearson correlation coefficient between per capita GDP and per capita EF for all countries with available data is 0.79. In other words, an average person in a rich country has much higher demands on nature's biologically productive resources than an average person in a poor country. Since some of these resources are globally shared (global climate and partly fisheries), rich countries consume a disproportionate share of global commons.[24] This leads us to the second interpretation of the EF concept.

We need to clearly explicate two concepts of bio-physical sustainability. It has been noted that using the EF concept for sustainability assessment we need to compare the EF of a population (typically of a country) with the available bio-capacity. One possibility is to compare a country's EF with the bio-capacity of the country's area (see Table 2, column 3). Under this concept, none of the five countries are sustainable. They have so low national bio-capacity (0.7 gha) that even with relatively low EF (1.1 gha) they show an ecological deficit (−0.4 gha), thus having higher demands on renewable resources than can be provided by the biologically productive areas of their respective territories. The second interpretation of sustainability could be called global egalitarian. Suppose all the people in the world, notwithstanding their countries of origin, have the same entitlement to the planet's natural resources. Then it makes sense to calculate average world bio-capacity per capita and compare it with national EFs per capita (see Table 2, column 5).[25] Here, East African countries show an ecological reserve, not only on the whole but also each of the five countries individually. There is a clear pattern here. According to EF, all five East African countries currently cannot live sustainably within their respective territories, while if they stake their claims on equal shares of global bio-capacity, their consumption is well within sustainable boundaries.

Whether these countries are on a sustainable trajectory thus crucially depends on the philosophical approach to sustainability. Reports by the Global Footprint

[24] While forests assumed to sequester carbon are within national jurisdictions, CO_2 emissions primarily affect the global atmosphere which is beyond national jurisdictions.

[25] This can also be applied to individuals. Rather than calculating EF as a national average it can be calculated on an individual level and compared to the average world bio-capacity.

Table 2 Ecological Footprint and balance (gha per capita, 2011)

	(1) Ecological Footprint	(2) National bio-capacity	(3) Ecological balance [2 – 1]	(4) Average world bio-capacity	(5) Ecological balance [4 – 1]
Burundi	0.8	0.3	−0.5	1.7	1.0
Kenya	1.0	0.5	−0.5	1.7	0.7
Rwanda	0.8	0.5	−0.3	1.7	0.9
Tanzania	1.2	1.0	−0.2	1.7	0.5
Uganda	1.2	0.6	−0.6	1.7	0.5
EA5	1.1	0.7	−0.4	1.7	0.6
SSA	1.2	1.4	0.2	1.7	0.6
LICs	1.0	1.1	0.2	1.7	0.7
MICs	1.9	1.6	−0.3	1.7	−0.1
HICs	5.1	3.0	−2.1	1.7	−3.4
World	2.7	1.7	−0.9	1.7	−0.9

Source: Global Footprint Network (2015); authors' calculations
Note: EA5: five East African countries, SSA: Sub-Saharan Africa, LICs: low income countries, MICs: middle income countries, HICs: high income countries. Total values for country groups are population-weighted averages. Numbers may not add up due to rounding off

Network (the only organization that calculates EF for most of the world's countries) show bio-capacity reserves/deficits based on local bio-capacity only. That is, sustainability of the EF concept is presented as living within the bio-capacity of the country's own territory. Given the unequal distribution of the world population and environmental resources, it is not very clear that this constitutes country sustainability. In fact, country (un)sustainability is negatively associated with population density, that is, countries with higher population densities tend to be less sustainable (more unsustainable).[26] While correlation does not mean causation, it is hard not to see any link here.

Rwanda has high population density with most of the bio-capacity coming from cropland, rather than from forests and fishing grounds. Going south from Rwanda to Namibia, these countries have very different population densities and ecological balance. With 452 people per sq km of land area, Rwanda shows an ecological deficit (−0.3 gha), while Namibia with less than 3 people per sq km shows an ecological reserve (4.4 gha). Yet, Rwanda's EF is three times lower than that of Namibia (0.8 versus 2.3 gha per capita). Supposedly, Rwanda cannot sustain its population on its territory without importing bio-capacity. Should not these countries be allowed to trade and offset their ecological balances? While such trade goes against the 'consumption within national bio-capacity' interpretation of EF, the same could be achieved by migration of people from Rwanda to Namibia.

[26] Pearson correlation coefficient for all countries with available data (after dropping three extreme outliers) is 0.42.

From a comparison of the six footprints on five related bio-capacities, we may gain additional insights into EA5 sustainability as measured by the EF concept. We leave out built-up land since the footprint always equals bio-capacity even at the local level. However, cropland and grazing land footprints closely follow related bio-capacities for each of the five countries, and the countries show very small deficits on fishing grounds (except Uganda). The real discrepancy between footprints and bio-capacity therefore lies in forests with a dominant share of the forest product footprint rather than the carbon footprint. While EF and ANS differ in many aspects, they both identify forests as a major area of unsustainability in East Africa (in ANS this applies to two countries, rather than all five). Yet, there are differences in how the two indicators approach sustainability.

EF is calculated for consumption and ANS for production. This distinction is important, as for example, CO_2 emissions embedded in production of a certain product are assigned to a country consuming that product in EF, while in ANS they belong to the producing country. In our case, while ANS tracks net loss of wood in the country (natural growth minus harvest), EF follows consumption of wood and compares it with natural growth of wood in the country (note that consumption does not equal production, but it is production plus import minus export). From a policy perspective, EF requires reduction in domestic consumption rather than halting domestic deforestation. Interestingly, sustainability as perceived by ANS also does not require countries to stop deforestation. Forest depletion is consistent with sustainability as long as all the resource rents are invested in physical capital (as GNS) or in human capital (as education expenditure). Since money is fungible, it is not easy to determine to what extent the rents have been invested. To achieve sustainability in both indicators, three approaches may be combined—reduction of wood consumption, reinvestment of resource rents and reduction of deforestation. The concrete policy mix may differ between countries, but given the importance of forests for eco-system services and the rate of deforestation in some East African countries (Burundi) in the recent past, it may be wise to give some importance to conservation of forests in these countries, though neither of the two indicators requires that.

5 Conclusion

This paper reviewed two composite indicators of sustainability and applied them to five East African countries. The indicators differ in their approach toward sustainability—EF focuses on natural renewable resources while ANS embraces both human-made and natural capital and allows for substitutability between them. It showed that these indicators have methodological and conceptual problems that limit their use as comprehensive aggregate measures of sustainability. Both include only limited types of environmental degradation and convert them into common units (money and global hectares) using contentious methodologies. While they can encompass more types of resource extraction and pollution, making them more comprehensive, the conversion methodology would be even less substantiated than the current version.

The results for EA5 countries show that only one country is unsustainable in ANS, while EF shows either all five countries as sustainable or all five as unsustainable, depending on the interpretation of bio-capacity. Since the dominant interpretation of the EF concept perceives sustainability as consumption within a regenerative capacity of a country's renewable resources, under which all EA5 countries are unsustainable, the two indicators provide a contradictory assessment of their sustainability. Yet, it is arguable whether a country's self-sufficiency within local biologically productive resources is a definition or precondition of sustainability in a global non-autarkic economy. One of the factors behind unsustainability may be high population density, as results from Rwanda and Burundi suggest. While high population density may indeed be a problem for sustainability, the way it is conceptualized by EF implies that a country will be more sustainable by simply reducing its population.

The indicators allow for disaggregation and identification of common sustainability issues. Both indicators identify forests as a major area of unsustainability in East Africa, but differ when it comes to the causes of that situation. While ANS see unsustainability in not investing all the resource rents from wood production (Burundi), for EF it originates in East Africa's consumption, whether the wood comes from domestic production or not. A sustainability policy may include reduction of wood consumption, reinvestment of resource rents and reduction of deforestation. Both indicators agree that CO_2 emissions are not a significant hindrance to sustainability in East Africa, but they disagree about that assessment on a global level—for EF, carbon emissions are the dominant factor behind unsustainability, but they are not of primary concern according to ANS. While the two methodologies are not directly comparable, it seems that CO_2 emissions are underestimated in ANS and overestimated in EF.

It may be unsettling that the indicators assess countries' sustainability differently and do not identify many common issues. Yet, given the different assumptions of the indicators, their methodological shortcomings and different views on what sustainability means at a country level, this is not so surprising. While none of the indicators may currently serve as a comprehensive aggregate measure of sustainability, it seems that ANS is able to offer a wider set of policy-relevant questions related to sustainability in East Africa and elsewhere. As for EF, we see less value in it as a metric of sustainability than as a tool that links consumers to the use of global environmental resources.

References

Arrow KJ, Dasgupta P, Goulder L, Mumford KJ, Oleson K (2012) Sustainability and the measurement of wealth. Environ Dev Econ 17:317–353

Bandura R (2008) A survey of composite indices measuring country performance: 2008 update. United Nations Development Programme, Office of Development Studies, New York, NY

Barrera-Roldán A, Saldívar-Valdés A (2002) Proposal and application of a Sustainable Development Index. Ecol Indicat 2(3):251–256

Bolt K, Matete M, Clemens M (2002) Manual for calculating adjusted net savings. The World Bank, Washington, DC

Borucke M, Moore D, Cranston G, Gracey K, Iha K, Larson J, Lazarus E, Morales JC, Wackernagel M, Galli A (2013) Accounting for demand and supply of the biosphere's regenerative capacity: the national footprint accounts' underlying methodology and framework. Ecol Indicat 24:518–533

Catton W (1980) Overshoot: the ecological basis of revolutionary change. University of Illinois Press, Urbana, IL, London

Costanza R, Daly HE (1992) Natural capital and sustainable development. Conserv Biol 6 (1):37–46

Daly H, Cobb J (1989) For the common good: redirecting the economy toward community, the environment, and a sustainable future. Beacon Press, Boston, MA

Ehrlich PR, Holdren JP (1972) A bulletin dialogue on "the closing circle": critique: one-dimensional ecology. Bull At Sci 28:16–27

Esty DC, Levy M, Srebotnjak T, de Sherbinin A (2005) 2005 Environmental Sustainability Index: benchmarking national environmental stewardship. Yale Center for Environmental Law & Policy, New Haven, CT

Eurostat (2012) Economy-wide material flow accounts (EW-MFA): compilation guide 2012. Eurostat, Luxembourg

Gajigo O, Mutambatsere E, Ndiaye G (2012) Royalty rates in African mining revisited: evidence from gold mining. Afr Econ Brief 3(6):1–12

Global Footprint Network (2015) National footprint accounts 2015 edition: http://www.footprintnetwork.org/en/index.php/GFN/page/footprint_data_and_results

Gupta S, Segura-Ubiergo A, Flores E (2014) Direct distribution of resource revenues: worth considering? IMF Staff Discussion Note SDN/14/05

Hamilton K (1994) Green adjustment to GDP. Resour Pol 20(3):155–168

Hardin G (1976) Carrying capacity as an ethical concept. Soundings 59:120–137

Hsu A et al. (2016) 2016 environmental performance index. Yale University, New Haven, CT, www.epi.yale.edu

Krausmann F, Erb KH, Gingrich S, Haberl H, Bondeau A, Gaube V, Lauk C, Plutzar C, Searchinger TD (2013) Global human appropriation of net primary production doubled in the 20th century. Proc Natl Acad Sci USA 110(25):10324–10329

Mederly P, Nováček P, Topercer J (2002) How to measure progress towards sustainability: the sustainable development index. Futures Res Q 18(2):5–24

Muganyizi TK (2012) Mining sector taxation in Tanzania. ICTD Research Report 1. Institute of Development Studies, Brighton

Neumayer E (2013) Weak versus strong sustainability: exploring the limits of two opposing paradigms, 4th edn. Edward Elgar, Cheltenham

Obua J, Agea JG, Ogwal JJ (2010) Status of forests in Uganda. Afr J Ecol 48(4):853–859

OECD (Organisation for Economic Co-operation and Development) (2008) Handbook on constructing composite indicators: methodology and user guide. OECD, Paris

PricewaterhouseCoopers (2013) The direct economic impact of gold. PricewaterhouseCoopers for the World Gold Council. http://www.gold.org/sites/default/files/documents/PWC_direct_economic_impact_of_gold.pdf

Ravallion M (2012) Mashup indices of development. World Bank Res Obs 27(1):1–32

Rees WE (1992) Ecological footprints and appropriated carrying capacity: what urban economics leaves out. Environ Urban 4(2):121–130

Solow RM (1991) Sustainability: an economist's perspective, the eighteenth J. Seward Johnson lecture to the Marine Policy Center, Woods Hole Oceanographic Institution. In: Dorfman R, Dorfman NS (eds) Economics of the environment: selected readings, 3rd edn. Norton, New York, NY, pp 179–187

Stevenson B, Wolfers J (2013) Subjective well-being and income: is there any evidence of satiation? Am Econ Rev Pap Proc 103(3):598–604

Stiglitz JE, Sen A, Fitoussi JP (2009) Report by the Commission on the measurement of economic performance and social progress. http://www.stiglitz-sen-fitoussi.fr/en/index.htm
Talberth J, Cobb C, Slattery N (2006) The Genuine Progress Indicator 2006: a tool for sustainable development. Redefining Progress, Oakland, CA
The Economist (2014) Tanzania's troubles over gas revenue: sharing the spoils. Retrieved from The Economist: http://www.economist.com/blogs/baobab/2014/09/tanzanias-troubles-over-gas-revenue
The World Bank (1997) Expanding the measure of wealth. The World Bank, Washington, DC
The World Bank (2011) The changing wealth of nations: measuring sustainable development in the new millennium. The World Bank, Washington, DC
The World Bank (2015a) World development indicators. http://data.worldbank.org/data-catalog/world-development-indicators
The World Bank (2015b) Doing business 2015: going beyond efficiency, 12th edn. The World Bank, Washington, DC
The World Bank (2015c) Worldwide governance indicators. http://info.worldbank.org/governance/wgi/index.aspx#home
van den Bergh JCJM, Verbruggen H (1999) Spatial sustainability, trade and indicators: an evaluation of the "ecological footprint". Ecol Econ 29(1):61–72
van den Bergh JCJM, Grazi F (2013) Ecological footprint policy? Land use as an environmental indicator. J Ind Ecol 18(1):10–19
van den Bergh JCJM, Grazi F (2014) Letter to the editor: response to Wackernagel. J Ind Ecol 18 (1):23–25
Vitousek PM, Ehrlich PR, Ehrlich AH, Matson PA (1986) Human appropriation of the products of photosynthesis. Biosci 36:368–373
Wackernagel M (2014) Letter to the editor: comment on "Ecological footprint policy? Land use as an environmental indicator". J Ind Ecol 18(1):20–23
Wackernagel M, Rees WE (1996) Our ecological footprint: reducing human impact on the earth. New Society Publishers, Gabriola Island
WCED (The World Commission on Environment and Development) (1987) Our common future. Oxford University Press, Oxford
WRI (World Resources Institute) (2015) CAIT 2.0: WRI's climate data explorer: http://www.wri.org/our-work/project/cait-climate-data-explorer

African Emergence, Inclusive and Sustainable Development and the Role of Social Science Research with Special Reference to Eastern Africa

Herman Musahara

Abstract This paper argues that the economic growth experience of several African countries and the Eastern Africa sub-region now generally constitutes an emergence. It is a necessary feature of a rising and shining Africa that has given cause to Afro optimism. Recovering from the scars of colonialism—subjugation, backwardness and darkness—and the post-colonial economic stagnation tagged as the 'African dummy', in the last 15 years Africa has witnessed the fastest economic growth, evidenced by credible cases. But it also provides evidence of rampant poverty, inequality and 'jobless' growth. Africa not only faces the challenge of inclusive development but also of how to make economic growth friendly for the environment and for sustaining resources required by future generations. This paper uses countries in Eastern Africa, including Ethiopia to demonstrate the challenges of sustaining the region's current economic emergence. It identifies the role of social science research in ensuring inclusive and sustainable development in the Eastern Africa sub-region.

Keywords African emergence • Social science research • Inclusive development • Sustainable development • Eastern Africa

1 Introduction

The argument of this paper is that the economic growth experience of several African countries and the Eastern Africa sub-region generally makes for an emergence. It is a necessary feature of a rising and shining Africa that has cause for Afro optimism. But this is not the first such feature and it is not sufficient. After the 1960s, there was the emergence of Africa out of colonialism—emergence from

H. Musahara (✉)
Organisation for Social Science Research in Eastern and Southern Africa (OSSREA), Addis Ababa, Ethiopia
e-mail: herman.musahara@gmail.com; musahara@ossrea.net

A. Heshmati (ed.), *Economic Integration, Currency Union, and Sustainable and Inclusive Growth in East Africa*, Advances in African Economic, Social and Political Development, DOI 10.1007/978-3-319-30432-8_10

subjugation, backwardness and darkness. However, it soon became clear that political emergence in the form of flag independence was not enough without economic change. The 'African dummy' (Barro 1991) was a euphemism for dysfunctional economies that seemed not to obey the behavior of production functions and growth models. The Harrod and Domar models died in Africa and the Solow model was not relevant. Africa got entangled in the economic downturn as structural adjustment either did not work or failed to transform African economies.

Now, African economies are emerging from long-term economic stagnation. The Solow and Kuznets models seem to be resurrecting after decades of trials and tribulations and what is emerging is what has been called the African developmental state. This is not derived from socialist versus capitalist economics. It is just a signification that what has been done by 'the Asian tigers' could also be done by the 'African lions'. A follow up to the argument is that this is also not enough. In the trajectory of development discourse, Africa and the world as a whole have to face the challenge of attaining inclusive and sustainable development. Data from Africa show that poverty and inequality are still rampant and pervasive no matter how economies are performing on gross domestic product (GDP) and GDP per capita. There are still worrying levels of unemployment. There is what has been called 'jobless growth'. The structures of the real sectors are still leaning on agriculture. And after all, as in Asia, it is not the whole of Africa as one country that is growing fast. It is some countries, may be about 30, with their own contexts, institutions and political memories. Several others are still mired in conflicts and soft states are not expected to meet the Millennium Development Goals (MDGs). But, as the African Union (AU) has indicated unequivocally, by 2050 and by Agenda 2063 they will work for a prosperous Africa based on inclusive and sustainable development (AUC 2013).

Africa is emerging and is economically promising. But there are also challenges of sustaining the gains. The first sets of challenges are largely in the distribution of the economic sphere itself, which still leaves pronounced poverty, inequality and unemployment. Africa and Asia still have the largest number of poor people. After Latin America, Africa has the second highest levels of both horizontal and vertical inequalities (Sachs 2015). There are challenges also in the spheres of environment, social inclusion and governance. The complete picture of emergence is thus a process that should go beyond economic growth to pro-poor economic growth with less inequalities or lower Gini coefficients. But, pro-poor economic growth also is not enough because here there are a lot of non-income inequalities. Given these realities, there is again a broader challenge of developing sound inclusive development strategies. The emergence should thus constitute resilient and green economies that can address the challenges of climate change and the hazards of the 'Anthropocene' (Sachs 2015). The emergence of Africa should not be seen in terms of lower Gini coefficients only but also in terms of the absence of all forms of exclusion—ethnic, religion or gender-based. Emergence of Africa should include efforts at mitigating conflicts and for promoting good governance, a lack of which will lead to poverty and erosion of development achievements. Emergence of Africa should involve sustainable development.

That said the less epistemic but operational argument emphasized in this paper is the role of social science research in emergence and inclusive sustainable development in the Eastern Africa sub-region. The paper argues that economic, behavioral, human and social sciences have a big role in the age of sustainable development that we are entering into.

This paper is a synthesis of works and presentations done by the author between 2014 and 2015. The first was editing a book on the Millennium Development Goals in Eastern and Southern Africa where the gaps in their implementation were noted. This work led to his attending and presenting a paper on the progress in implementing MDGs at an international conference convened in Rwanda in early 2015. Subsequently the author attended and presented a paper on the 'Emergence of Africa and the Development State' at a conference in Abidjan convened by UNDP, the World Bank and the African Development Bank. He also presented a keynote address at an international conference convened for Eastern African universities by the University of Rwanda and Jonkoping University in March 2015. The speech was on inclusive and sustainable development. Most of the data are derived from the World Development Indicators (WDI 2015) and the Human Development Report (2014).

The rest of the paper is organized in four sections. Emergence of Africa section describes the notions of emergence of Africa—emergence, inclusive development and sustainable development. Challenges of Inclusive and Sustainable Development section gives an empirical view of the challenges of inclusive development. The Roles of Social Science Research section outlines the roles of social science research in informing the emergence of Africa as a holistic and multi-disciplinary process. Conclusion section presents conclusions for the argument.

2 Emergence of Africa

Pre-colonial and colonial Africa was a dark continent, full of disease, animals and the backwardness of its indigenous people. A significant event for twentieth century Africa and Africans was decolonization, which enabled most African countries to emerge from colonialism in the 1960s—they emerged from slavery, bondage and exploitation and joined the world of free nation-states. The 1960s through 1970s saw a lot of political and cultural Africanism and pan Africanism. But, challenges started showing not so long after the decolonization. Scores of African countries started being identified with dictatorships, coup d'états, conflict, clientelism and corruption.

Starting with the 1973 oil crisis, also known as the 1973 Oil Embargo, African economies started facing economic problems stemming mainly from imported inflation. Balance of payments problems started making most African economies highly indebted. Multilateral financial institutions, more notably the International

Monetary Fund (IMF), saw the logic for reforms of most of the economies. With more shocks in this decade and through the 1980s, most economies succumbed to structural adjustment programs (SAPs) administered by IMF. State subsidized programs were removed, economic and financial activities liberalized and industries privatized. However, SAPs and reliance on market-led reforms were more harmful than helpful in many of the reforming economies. Devaluation of currencies as the most common prescription did not provide the needed cure (Musahara 2004). Thus, the 1990s were marked by the introduction of state-administered Poverty Reduction Strategy Papers (PRSPs). Though these were criticized as well, many African economies started considering development as an ideology and looking at what was happening elsewhere in South East Asia. Ever since, the economic performance of many countries has shown patterns that are promising and different from what had come to be called in development jargon the 'lost decades' of the 1970s and 1980s. An introduction to a recent international conference on 'Emergence of Africa', held in March 2015 in Abidjan, Ivory Coast, had the following preamble:

> There is a sharp contrast between the current economic dynamic in Africa and that of the 1980s characterized by a stagnant growth and over-indebtedness. Africa currently ranks among the world's regions with strong economic growth, despite the global economic recession triggered by the financial crisis of 2008. Africa's economic performance is such that the current debate is focused on transforming Africa from a low-income continent to a middle-income continent. This change of focus is justified by an average growth rate of about 5% over the past decade, while developed countries, including the US and the European Union experienced a long recession over the same period (http://www.africaemergence.com/actualite.php?lang=en&ID=1).

An Article in *The Economist* in December 2011 'Africa Rising' has often been referred to as the concession of western scholarship to a process of an emerging Africa. It states aptly that 'Despite Africa's long history of corruption, violence, famine and disease, the continent is on the economic rise'. It notes that over the past decade, 6 of the world's 10 fastest-growing countries were in Africa. *The Economist* (3 December) cites the following factors (sic):

1. *The commodities market is booming*: Between 2000 and 2008, approximately one quarter of Africa's revenues came from natural resources.
2. *Population is growing*: As fertility rates are crashing in Asia and Latin America, Africa's population is projected to grow substantially over the next 40 years.
3. *Africa has a fast-growing middle class*: According to Standard Bank, around 60 million Africans earn $3000 a year, and 100 million people will earn that much by 2015.
4. *Cross-border commerce is growing*: Tariffs and trade barriers are falling between African nations. In addition, many political rivalries are fading.
5. *Africa has more than 600 million mobile phone users—more than America or Europe*: Approximately one-tenth of Africa's land is connected by mobile-internet services.

Along with these positive economic indicators, many Africans have a much better quality of life and future outlook:

1. Africans are healthier now due to wider distribution of mosquito nets and a decrease in HIV/AIDS.
2. Many more Africans of working age are entering the job market with higher levels of education.
3. Previously explosive birth rates are beginning to decline.
4. Most African nations are at peace and operating with functional governments.

Overall, the African economy grew by 6 % in 2011 and nearly 6 % in 2012—about the same as Asia. Other than *The Economist*, a UN Report (UN 2015) indicates that GDP growth in Africa is expected to accelerate to 4.6 % in 2015 and 4.9 % in 2016. Other estimates are even more ambitious. *The World Economic Situation and Prospects (2015)* estimates that the least developed countries (LDCs) in Africa will continue to exceed the global average growth of GDP with expected acceleration from 5.3 % in 2014 to 5.7 % in 2015, mainly due to an improving external environment. Foreign direct investment (FDI) was US$ 110 billion in 1998 which rose to US$ 554 billion in 2010 (ACET 2014). If Africa were a country, it would, with a per capita of US$ 1500, classify as a middle income country. By 2050, Africa will have a larger and younger workforce than China or India. It will continue to transform its economies through exploiting land and natural resources, economic diversification, enhancing competitiveness, increasing the share of manufacturing, use of technology and reducing dependency on foreign assistance. The big question here is how sustainable is the Afro optimism? In the next section, the paper argues that Africa now has to identify challenges of inclusive and sustainable development that it needs to address.

3 Challenges of Inclusive and Sustainable Development

The emergence of Africa is a concept that evokes the need for inclusive and sustainable development as yet another paradigm shift in the meaning of development. Six decades on, development is more than GDP and GDP per capita. If the emergence of Africa is seen using the lenses of Harrod (1939) and Domar (1946), then it is good and necessary but not sufficient. As with Schumpeter (1961), growth is part of development. For other scholars, development was a process in stages (Rostow 1960). For Kuznets, what is being experienced is what can be generically seen as the U curve of development and for Solow, this a convergence trend wherein poor countries experience faster rates of economic growth.

Interrogating the limits of economic growth alone started in the 1980s. For example, in the 1980s emerged the Basic Needs approach (Streeten 1979). The defining features of development include how individuals, households and societies had access to food, shelter and clothing as indicators of their development. By 1990 the concept of human development had taken root and soon the Human

Development Index was launched (Sen 1988; UNDP 1990). The concept of development included, of course, GDP per capita, but also quality of life in terms of life expectancy and schooling. It included development as choice, as entitlements, as capabilities and as freedom (Sen 1999).

It is worth noting that the concept of sustainability was introduced and gave way to the more robust concept of sustainable development at a later date. In the 1970s, sustainability referred to an economy in equilibrium with a basic ecological support system (Stivers 1976). Sustainable development is understood as an operable definition of the principle of sustainability (Cornwall and Eade 2013). The concept of sustainable development is being used by the United Nations since the Brandt Commission (1987) defined sustainable development as 'development that meets the needs of the present without compromising the ability of future generations' (WCED 1987: 8). It became even more popular in 1992 at the World Summit on Environment in Rio de Janeiro. Another global summit on sustainable development took place in Durban in 2002 and was the origin of the jargon Rio+ 20. Today, there is almost a consensus that the current development model has been inadequate. Sustainable development includes not only economic development but also social, environmental and governance issues (UN 2013). It is the new approach to sustainability that can face multiple problems of poverty, inequality and inequity, food security, effects of climate change, social exclusion and gender, environmental degradation, cities, energy needs, water crisis and migration to mention but a few (WSSC 2013). In September 2015, world leaders are expected to endorse Sustainable Development Goals (SDGs) as a successor to the just-ending MDGs. It has been noted that we are entering the Age of Sustainable Development (Sachs 2015).

In the effort at searching for inclusive development, some of the challenges that we are identifying today also started being noted. There was, for example, the debate on redistribution and growth (Alesina and Rodrick 1994; Chenery 1991; Dollar and Kraay 2004; The World Bank 2000). While initially the debates were around the strengths and weaknesses of growth with redistribution, the real issues that emerged were that poverty, inequality and unemployment were critical economic concerns.

At that juncture, the concept of pro-poor and inclusive growth started becoming relevant in the development discourse. But, even today, these concepts are used easily and interchangeably although they are quite distinct (Kanbur and Rauniyar 2010). When poverty is reduced substantially compared to the rate of growth of say GDP per capita, the phenomenon is called pro-poor growth. But the problem that arises which is useful for the emergence of Africa is where the growth of the economy has led to poverty reduction but has also resulted in high levels of inequalities in income and consumption but also horizontally and vertically. This means, growth can be pro-poor without being inclusive. In fact, there can be growth with poverty reduction while there are segments of the poor that have become worse in the process. The definition and debates on pro-poor growth occupied scholars for the most part of the 2000s (Kakwani and Pernia 2000; Dollar and Kraay 2004; OECD 2005; Ravallion and Chen 2001; UNDP 2000; White and

Anderson 2001). At a more elementary level let us make a clear analysis of which pro-poor growth is referred to in this article.

It can be argued that the basis for inclusive and sustainable development is pro-poor growth. Pro-poor economic growth will reduce poverty. But for the reduction of poverty to be sustained and have a lasting impact on development, the entire pattern of development has to be influenced (OECD 2005). As argued earlier, pro-poor growth is also necessary but not sufficient. It is not enough to know that economic growth is to the benefit of the poor without information on how to implement the pro-poor policy approach. In emerging economies of Africa it is important to distinguish between growth that benefits the poor and other growth. It is important to see whether growth in incomes of the poor is disproportionately higher than growth in incomes of the non-poor (OECD 2005). Three distinct patterns may be conceivable as a basis of inclusive development.

There is weak pro-poor growth if the income growth rates of the poor are more than zero. There is relative pro-poor growth if the income growth of the poor is higher than the average growth and relative inequality falls. Finally, there will be the most desirable form of strong pro-poor growth where the income growth of the poor is strong and absolute inequalities fall. It can be established by a simple inspection of studies which shows that there has been a generalized interest in the first two forms (OECD 2005). But reduction of absolute inequality is the most important in planning for inclusive growth. Addressing the issue of strong pro-poor growth is important but as said it is not enough as there are a number of non-income aspects of poverty which are not considered. As noted earlier there are also qualitative aspects of development such as gender empowerment and rights protection that characterize the overall meaning of inclusive development. In East Africa it is not adequate to characterize the emergence of the countries on the basis of their economic growth and performance over the last 15 years alone.

Thus, the concept of inclusive growth has been coined to indicate growth that has both reduced poverty and is accompanied by lower income inequalities. This defines inclusive development as referring to improvements in distribution of well-being along dimensions beyond growth, at the same time as an average improvement in achievements is realized. Development includes economic growth, changes in the structure of production, changes in the spatial distribution of the population such as urbanization and improvements in social indicators.

Theoretical distinctions are important when one looks at the African experience now. Recently, about 23 countries in Sub-Saharan Africa graduated to middle income countries (see Fig. 1) (Norton and Rogerson 2012). Despite good growth performance, poverty reduction is still sluggish and the manufacturing sector is declining (JICA 2013) just as the concept of 'jobless growth' means a weak job market and increasing inequalities. These are indications that development has not been inclusive because growth has not been inclusive. Inclusive growth has been defined as a reduction in disparities beyond tolerable thresholds and everybody in society is given an opportunity to take part in productive processes. Sachs (2004) notes that inclusiveness also means a form of development which encompasses civil, civic and political rights. Norton and Rogerson (2012) note that inclusiveness

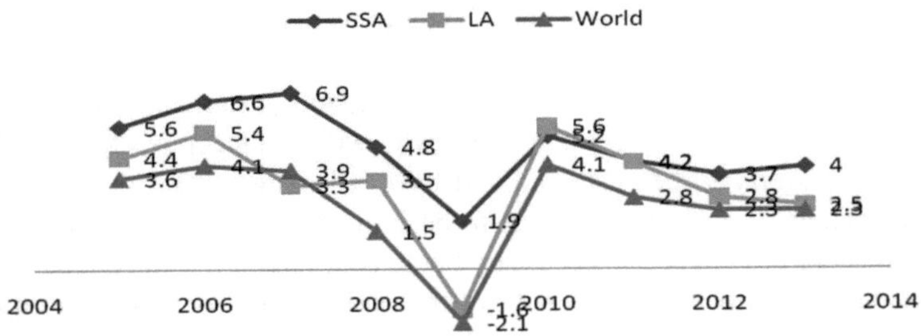

Fig. 1 Recent Sub-Saharan African growth trends seen against Latin American and world trends. *Source*: Drawn by the author using WDI Data

means poverty reduction and also rights. Power and empowerment need also to come out as objectives of development. While development has been seen as non-inclusive because of lack of opportunities, there is still evidence of social exclusion related to discrimination by age, religion, gender, disability, poverty and ethnicity (UNDP 2014). Inclusive development has also been expanded to include human rights, participation, non-discrimination and accountability (UNDP 2014).[1]

Norton and Rogerson (2012) further note that besides growth and poverty reduction the challenges of inclusive development include climate change and violence and conflict. While climate change is a global challenge, the poor in Africa are the most vulnerable to its impacts. Floods and droughts affect agriculture which is the mainstay of a majority of the poor in Africa. Some mitigation efforts such as use of bio-fuels affect the poor by interfering with agricultural production. Although conflict in Africa dropped dramatically in Sub-Saharan Africa (SSA) (Fig. 1) in the last decade, fragile states still harbor a majority of the poor.

Institutions for inclusive development are also weak and not working properly. The roles of the state, private sector, aid and other stakeholders need to be transformed (OSSREA 2004). Oxfam (2014) talks of 'ensuring benefits for all'. Inclusive development is driven by people through open dialogues between country governments and their citizens. Inclusive development is a pro-poor approach that equally values and incorporates contributions from all stakeholders, including marginalized groups, in addressing development issues. It promotes transparency and accountability and enhances development cooperation outcomes through collaboration between civil society, governments and private sector actors.

As has been noted the concept of the emergence of Africa will make meaningful change and also a shift in its global position if economic growth leads to inclusive and sustainable development. Let us now look in general at these indicators in countries of East Africa, including Ethiopia. Of course, just like Africa as a whole, countries and regions have their own contexts and contents.

[1] http://www.undp.org/content/undp/en/home/ourwork/povertyreduction/focus_areas/focus_inclusive_development

Table 1 Annual growth rates of GDP in Eastern Africa

	2000	2001	2002	2003	2004	2005	2006	2007	2008	2009	2010	2011	2012	2013
Kenya	0.6	3.8	0.5	2.9	5.1	5.9	6.3	7	–	3.3	8.4	6.1	4.5	5.7
Uganda	3.1	5.2	8.7	6.5	6.8	6.3	10.8	8.4	8.7	7.3	5.9	6.6	3.4	6.0
Tanzania	4.9	6.0	7.2	6.9	7.8	7.4	6.7	7.1	7.4	6.0	7.0	6.4	6.9	7.0
Rwanda	8.3	8.7	13.5	1.5	6.9	6.9	9.2	7.6	11.2	6.3	7.3	7.9	8.8	4.7
Burundi						1.0	5.4	4.8	5.0	3.5	3.8	4.2	4.0	4.6
Ethiopia	6.1	8.3	1.5	−2.2	13.6	11.8	10.8	11.5	10.8	8.8	12.6	11.2	8.6	10.5

Source: WDI (2015)

Table 2 Structure of growth in some African countries

	Agriculture	Agriculture	Industry	Industry	Services	Services
	2000	2012	2000	2012	2000	2012
Kenya	32	29	17	21	51	50
Uganda	29	26	23	29	48	46
Tanzania	33	29	19	24	47	47
Rwanda	57	33	14	14	49	52
Burundi	NA	NA	NA	NA	NA	NA
Ethiopia	48	48	12	10	40	42

Source: WDI (2015)

A scan of Table 1 shows that generally in Africa growth has been positive. The noticeable negative growth in the last 15 years was in Ethiopia in 2003. Yet, Ethiopia and Rwanda are the two countries that have consistently registered positive growth averages in their economies, making them two of the fast growing economies in East Africa. Uganda also has on average an above 5 % rate of growth (Table 2). This is consistent with the overall growth episode of Sub-Saharan Africa mentioned earlier.

A partial explanation for this growth that has not been pro-poor and thus not inclusive is in the sources of the growth. Table 2 shows that a major contribution to GDP is services while it is known that a majority of the poor reside in rural areas. Ideally, an economy undergoing structural transformation will see a bigger contribution coming from industry and preferably from manufacturing (Sutcliffe 1980). This does not seem to be the case, and even if it did, it would be useful for inclusive growth if it created more jobs. The question of sources of growth also questions the question of sustainability of growth and the issue of how the growth is transforming the economy. It also implicitly points to low agricultural productivity levels still rampant in most SSA countries.

It has also been argued that growth has not been inclusive enough because there is still rampant poverty in these countries. Despite notable growth rates in most of Eastern and Southern African countries, poverty by different measures is still in high percentages in double digits (Table 3) Despite impressive growth, the countries in the region are still low on the Human Development Index, with only Kenya and Rwanda having slightly higher than 0.5 on the Index where countries score between 0 and 1 (Table 4).

It has been indicated that the economies of East African countries are not yet inclusive enough because there is still rampant poverty. With the exception of Kenya, more than 50 % of the population of East Africa was under multiple forms of poverty in 2014. Using the percentage of people in poverty under less than US$ 1.25 a day, only Uganda and Ethiopia are below the Sub-Saharan average rate of 40 %. Economic growth cannot lead to durable development if poverty is still such a big challenge.

Another indicator of non-inclusive development is levels of inequalities. Table 5 shows incomes held by the highest and lowest 10 % of the population in Eastern Africa. Data available is mostly for 2010 and 2011. With the exception of Ethiopia,

Table 3 Poverty by head count index of less than US$ 1.25 a day between 2009 and 2011

	Country	Poverty level by percentage
1	Kenya	43.7
2	Uganda	38.0
3	Tanzania	67.9
4	Rwanda	63.2
5	Burundi	81.3
6	Ethiopia	30.7
7	Africa	40.0

Source: WDI (2015)

Table 4 The most current socioeconomic indicators of East African economies

Country	HDI	Rank	LE	MYRS	PPP	GII	PIMP	E2PR	Total population
Kenya	0.535	147	61.7	6.3	2158	0.55	48.2	75.6	44.4
Uganda	0.484	164	59.2	5.4	1335	0.53	70.3	86.8	37.6
Tanzania	0.485	159	61.5	5.1	1702	0.55	63.4	91.9	49.3
Rwanda	0.506	151	64.1	3.4	1403	0.41	70.7	92.6	11.8
Burundi	0389	180	54.1	2.6	749	0.50	81.8	87.8	10.2
Ethiopia	0435	173	63.6	2.4	1303	0.55	88.3	84.1	94.1

Source: HDR (2014)
Note: HDI = Human Development Index, LE = Life Expectancy, MYRS = Mean Years of Schooling, PPP = Purchasing Power Parity, GII = Gender Inequality Index, PIMP = People Under Multiple Poverty, E2PR = Employment to Population Ratio

which had 27.5 % of income held by the highest 10 %, the rest the highest 10 % of the population held more than 30 % of the income.

Among the measures of levels of disparities or inequalities in categories of people is gender. From Table 6, it can be seen that only Rwanda has a high gender equality rate of 4 on a scale of 1–6. Others are marginally over the average with 3.5. Economies doing very well like Ethiopia just score an average.

From the last column of Table 7, another more common measure of inequality is the Gini coefficient. During 2003–2013 the average Gini coefficient was higher than 0.40 % which is a relatively high level of inequality.

Sustainability is mainly about considering the environment in matters of promoting economic growth. From Table 8 it can be seen that only Uganda scores 4 on a scale of 1–6 in terms of countries that consider environment in their policies.

Unemployment is also a challenge of inclusive and sustainable development. From Fig. 2, it can be seen that Tanzania had the highest levels of unemployment while Rwanda had the lowest rates. The real situation, however, depended on the type of measures used. For example, today we talk of the 'working poor', which means people who are numerically employed but who have jobs that are not decent enough to provide a decent living.

So, generally as stated by AfDB, Eastern Africa, like the rest of SSA is undergoing an 'exciting economic renaissance'. For the last 10 years, the GDP growth average has been 6.6 % and annual real growth rate of countries in the East

Table 5 Levels of inequalities

Country	Income held by highest 10pc	Income held lowest by 20pc
Ethiopia	27.5 (2011)	8.0 (2011)
Rwanda	43.2 (2011)	5.2 (2011)
Tanzania	31.1 (2012)	7.4 (2012)
Uganda	35.8 (2013)	5.8 (2013)

Source: UNDP (2014)

Table 6 Gender (gender equality rate Low =1 High =6)

Country	2010	2011	2012	2013
Kenya	3.5	3.5	3.5	3.5
Uganda	3.5	3.5	3.5	3.5
Tanzania	3.5	3.5	3.5	3.5
Rwanda	4	4	4	4
Ethiopia	3	3	3	3

Source: UNDP (2014)

Table 7 Human development ranking for Eastern and Southern Africa

Country	Category of HDI	HDI rank 2013	HDI change 2008–2013	GDI rank	Gini 2003–2013
Kenya	Low	147	−1	107	47.7
Uganda	Low	164	−4	114	44.3
Tanzania	Low	159	NA	100	37.6
Rwanda	Low	151	17	80	50.8
Ethiopia	Low	173	2	126	33.6

Source: UNDP (2014), Human Development Report (1990, 2014)

Table 8 Environment

Country	2010	2011	2012	2013
Kenya	3.5	3.5	3.5	3.
Uganda	4	3.5	3.5	3.5
Tanzania	3.5	3.5	3.5	3.5
Rwanda	3.5	3.5	3.5	3.5
Ethiopia	3	3.5	3.5	3.5

Low =1, high = 6
Source: UNDP (2014)

African region has tended to converge around 4.5 %. East Africa has a youthful population as well with 70 % under the age of 30 years. East Africa is discovering a lot of natural resources including oil and gas. We can also say that East Africa is also rising and shining. ICT has changed the structure of the economy mainly by taking over agriculture to contribute about 42 % to GDP. Rwanda has improved its business environment tremendously and has shot high among the top countries of the world in the 'Doing Business' rankings.

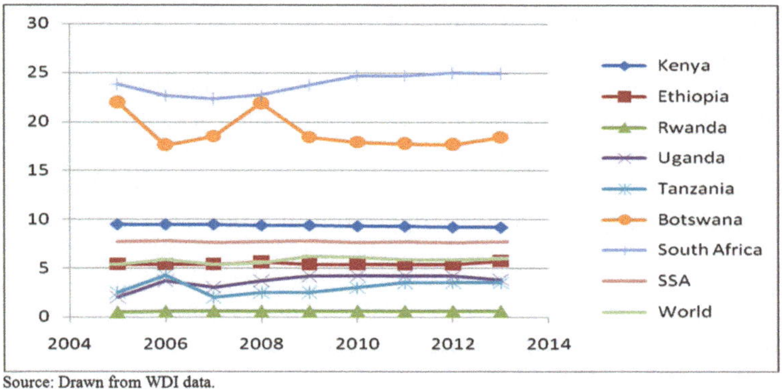

Fig. 2 Unemployment. *Source*: Drawn from WDI data

But considering the region as a whole and each country individually, every scholar concedes that the growth has not been inclusive. Burundi presents an example where not only high levels of poverty challenge inclusive and sustainable development but also the impact of good GDP growth rates gets easily dampened by political instability. Burundi also has limited arable land and a majority of the poor—almost 88 %—live in rural areas compared to 12 % who live in urban areas. This contrast means that there is a considerable amount of disparity or inequality among different classes of the poor.

Ethiopia has one of the largest economies in Africa and a population in excess of 90 million. It has experienced very good growth with real GDP growing at an average of 11 % in the past 15 years. But, like in the rest of the region, the growth has not managed to employ the available labor force and urban cities have not become the drivers of growth.

Kenya, as noted, has managed to shift its economy and depends more on services. More than 60 % of its GDP comes from services and about 16 % comes from industry. This is not an adequate structural transformation as expansion of manufacturing will be more contributive to more durable development. The Gini coefficient has been under 0.5 but quite high. The explanation of inequality in Kenya is attributed to past neglect of some regions of the country. It is estimated that 12.7 % of the youth are unemployed.

Rwanda is another success story on the growth front. Despite suspension of aid, Rwanda has continued to realize growth rates of GDP above 5 %. Between 2001 and 2013, industry grew faster than agriculture, but the economy was still dependent on agriculture. Rwanda had the best improvement in gender equality and equity. However, although Rwanda had the lowest rate of recorded unemployment, it was believed that 42 % of its youth were underemployed.

Tanzania was noted to be one of the fastest growing economies in Africa with an average of 6.9 % rate of growth between 2010 and 2013. It discovered deposits of natural gas and got a lot of its growth from trade. But Tanzania has what is called notable basic needs poverty that, though going down, is still high. The poverty rate was 28.2 % in 2012 while the rate was 33 % in 2007. Tanzania has also not undergone structural transformation and depends on agriculture.

Uganda has also discovered natural resources and, although investing in oil, its contribution to the economy has continued to be too little at less than 0.5 %. Poverty has declined in Rwanda but not as much as the economy has grown, that is, it has been less inclusive as well. For inclusive growth, there is still need to make some legislations that strengthen women's rights to land assets and inheritance.

4 The Roles of Social Science Research

It was noted in the previous sections that two major challenges of African emergence are how to ensure inclusive growth and sustainable development. This is echoed clearly by the African Vision as articulated in Agenda 2063 which states that Africa needs to 'plan for a prosperous region based on inclusive and sustainable development.' In this sense, the problem is not in the diagnosis of the challenges for policymakers, but in debunking the challenges that are gradually becoming visible at both academic and policy levels. In this section we use two levels to indicate the role of social science research for emergence and inclusive sustainable growth in Africa in general and in Eastern Africa in particular.

Firstly, in the academic sphere and at the global level, the World Social Science Report (2013) identified the role of social science research in a changing global environment—that of playing a role in achieving sustainable development. Secondly, 2015 marks the end of the MDGs. The gaps in the goals and in their achievements have informed the drafting of SDGs. It is becoming clearer that the cross-cutting deficit of the MDGs has been inadequate inclusive development. Binding the two challenges of inclusiveness and attention to the environment and climate change is the fact that the period from 2016 has been named the Age of Sustainable Development (Sachs 2015).

The first point for social science research is that it has to contribute to an understanding of 'human behavior' and how it can be changed to foster inclusive and sustainable development. The four elements of sustainable development are economic, social inclusion, environment and governance. Thus, while poverty reduction should provide the basis of inclusive growth, social inclusion also means reduction of discrimination by gender, age, ethnicity, religion and other distinctions. The importance of the environment has come to be dominated by the lobby on the effects of climate change. Social sciences in the age of sustainable development need to demystify the effect of climate change. The World Social Science Report (2013) is concerned with the focus on the challenges of understanding the impact of climate change.

In this report social science research is viewed in terms of contributing to the process of 'transformations to sustainability' (Hackman and St Clair 2012). Six different cornerstones are suggested by the authors as areas where social science research can play a role.

Firstly, social science research needs to assist in an understanding of the political economy of the processes of global change. There is also the role of clarifying the interdependence of people's vulnerabilities and changes. Social science research will provide an understanding of the context of global change risks, impacts, perceptions, experiences and responses and the differences by social classes, gender, race or faith and other categories. This cornerstone is important generically to inclusive development.

Secondly, social science research needs to identify and map current and future threats from global change. The areas include impact on people and community coping mechanisms, responses, innovations and boundaries. The current performance of African economies should be cognizant of these possible impacts.

Thirdly, social science research is also concerned with what drives individual and collective social change. There are also best practices from elsewhere and how these can be scaled up. Social science research can provide evidence on socially desirable changes, lifestyles and an alternative future.

Fourth, is the issue of interpretation and making sense. This includes values, beliefs, interests, worldviews, hope, needs and desires behind people's experiences and global change.

The fifth cornerstone is responsibilities and the research areas include obligations, duties and responsibilities towards the poor and vulnerable and also for future generations. Social science should help in identifying the ethical dimensions of the interpretation.

Sixthly, social science research is also about governance and decision making. It should assist in knowing how decisions are made under uncertainty and pathways available for decision making.

Sachs (2015) notes four areas where social sciences are important for sustainable development research. First, is epistemic work on areas such as poverty, hunger and climate change and access to health, education and food security. Second, are new research and demonstration programs. These are programs for promoting specific innovations needed for sustainable development. Third, is an improved understanding and design of global social, economic and technological changes such as eradication of poverty, or heading environmental catastrophes such as tools of advocacy, analysing ways of addressing delayed action against hazards of climate change and designing sustainable development goals and finally on how social scientists organize themselves in view of the need for providing solutions to the threats that humanity is facing.

The best summary of the role of social sciences is by Irina Bokova (2013), who while introducing the WSSC report says, 'human activity is the major force shaping the planetary system...we shape our environment as it shapes us...social sciences has the role to contribute to social transformation' and then WSSC (2013) points out correctly that human behavior is important in understanding and averting the global crisis and social sciences are uniquely positioned to help shift the current development paradigm.

The second side of the role of social science research more specifically deals with covering the gaps identified in the implementation of MDGs and provides

Table 9 Gaps in implementation of millennium goals in Eastern Africa

	The Goal	Countries that have done well	Countries that were not the best performers
1	Eradicate extreme poverty and hunger	Ethiopia, especially in target 1 B	Rwanda, Tanzania, Kenya, Uganda, Burundi
2	Achieve universal primary education	Rwanda and Tanzania	Kenya, Burundi, Ethiopia and Uganda
3	Promote gender equality and empower women	Rwanda and Ethiopia	Kenya, Burundi, Uganda and Tanzania
4	Reduce child mortality	Rwanda	Uganda, Kenya, Tanzania, Ethiopia, Burundi
5	Improve maternal health	Rwanda	Uganda, Kenya, Tanzania, Ethiopia, Burundi
6	Combat HIV/AIDS, TB, malaria and other diseases	Rwanda	Uganda, Kenya, Tanzania, Ethiopia, Burundi
7	Ensure environmental sustainability	Ethiopia	Uganda, Kenya, Tanzania, Rwanda, Burundi
8	Global partnership for development	Rwanda	Uganda, Kenya, Tanzania, Ethiopia, Burundi

Source: UN (2015)

evidence about the lack of good performances. Using the Eastern Africa case there is a need for a deep analysis on why some countries did not do as well as the rest and how the particular goals can be addressed in the post MDGs period (Table 9).

As mentioned earlier not performing well in meeting MDGs does not mean that this is the only reason why we need social science research because there are also gaps that are not related to implementation but to how the goals were designed. For instance, every goal is definitely related to poverty reduction (MDG 1). Secondly, gender is a cross-cutting issue which has been put in just one goal—MDG 3. But above all there are other aspects of poverty and aspects of the goals that do not appear anywhere. These will appear in the SDGs and as Sachs points out social science research needs to provide pathways in implementing the SDGs when they are adopted in September 2015 (Table 10).

5 Conclusion

Emergence of Africa is good news after the lost decades and African dummy variables of the 1980s. More than half the countries in Africa have experienced considerably high rates of growth over the last 15 years. It is Africa rising, Afro optimism and Afro enthusiasm. This paper has argued that despite this impressive performance we need to keep in mind several challenges.

Poverty is still rampant, inequality indicators are high and jobs are not as forthcoming as was expected of the growth. The African economies are rising but they have not structurally transformed. A lot of growth is still derived from a low

Table 10 Research areas in social sciences for supporting inclusive and sustainable development

Social science research area	Remark
Poverty is still a major problem	Analyse why it has not been possible to reduce poverty substantially and how to address it under SDGs
Dynamics and drivers of inequality	Major challenges to inclusive development
Non-income inequalities and inequities	Major challenges to inclusive development
Employment and jobless growth dynamics	Major challenges to inclusive development
Quality of education	Access to, availability, financing, relevance models
Urbanization	More on closing rural and urban development gaps as an inclusive development issue but also smart cities as sustainable development
Infrastructure	Did not appear in MDGs but is a big driver of development
Good governance and security	Part of a new definition of sustainable development and did not appear under MDGs
Gender studies	Social inclusion
Climate change metrics and understanding	Sustainable developments
Human rights	Area did not appear under MDGs
Social protection	Important for inclusive development
Methodologies	Methodologies for best and credible analysis

Source: WSSC Report (2013)

productive agricultural sector. African growth needs to be pro-poor. However, being pro-poor is seen to be inadequate. Growth also has to be inclusive to reduce both poverty and inequalities. This paper also argued that the work done here has also been inadequate. An inclusive development strategy must not only be about growth but also about other non-income inequality reduction, expansion of opportunities and their access and setting up frameworks that promote participation of all in the economic growth process. If these are in place, then it will be possible to move towards inclusive development.

Inclusive development is in turn part of a broader concept of sustainable development. Not only should countries emerging from stagnation in Africa address threats to the environment and the adverse effects of climate change, in addition to non-income inequalities they should also address issues of social exclusion including gender inequalities and bad governance. Conflict is still a visible cause of poverty in several parts of Africa.

In searching for sustainable development, research in social sciences has to play its role. Evidence, knowledge products, information and policy analyses for change are required. It is thus safe to conclude that while the apparent rising of Africa in general and of Eastern Africa in particular is cause for celebration more needs to be done to sustain economic growth, to change the structure of the economies, to promote inclusive development beyond pro-poor growth and economic indices to include social inclusion, environmental considerations including climate change adaptation and mitigation as well as promotion of good governance. This being a tall order policy and academic agendas need evidence and policy analyses for ensuring that the nascent African development becomes robust and durable.

References

ACET (2014) Growth with depth. African Transformation Report 2014. ACET, Accra

Alesina A, Rodrick D (1994) Distributive politics and Economic growth. Q J Econ Res 109 (2):465–490

AUC (2013) Agenda 2063. Africa we want. Planning division. African Union. Addis Ababa

Barro R (1991) The quest for African dummy. Economic history. London School of Economics, London

Bokova I (2013) World social science report. International Social Science Council. UNESCO, Paris

Chenery H (1991) Redistribution with growth. World Bank, Washington, DC

Cornwall A, Eade D (eds) (2013) Deconstructing development discourse. Buzzwords and fuzzwods. Practical Action Publishing. Institute of Development Studies, Sussex

Domar E (1946) Capital expansion, growth and employment. Econometrica 14(2):137–147

Dollar E, Kraay D (2004) Growth is good for the poor. The World Bank, Washington, DC

Hackman H, St Clair AL (2012) Transformations to sustainability. WSSC, Paris

Harrod R (1939) An essay in dynamic theory. Econ J 193:14–33

JICA (2013) For inclusive and dynamic development in Sub-Saharan Africa. Jica Research Institute Report, Tokyo

Kakwani N, Pernia P (2000) What is pro-poor? Policy innovations. www.policyinnovations.org

Kanbur R, Rauniyar R (2010) Inclusive development: two papers on conceptualization and application. Asian Development Bank, Mandaluyong, Philippines

Musahara H (2004) Poverty and Government expenditure. The effect of government expenditure on poor groups with reference to Rwanda. PhD Thesis, University of Western Cape

OECD (2005) Promoting pro-poor growth. Key policy message. Guidance for donors. Paris

OSSREA (2004) Achieving inclusive development in Africa. Seminar sponsored by Dutch Government for Knowledge Platform on Development Policies and ODI. Addis Ababa

Oxfam (2014) What is inclusive development? Briefing note. Oxfam, UK

Norton A, Rogerson A (2012) Inclusive and sustainable development. Paper prepared for International High Level conference to mark 50th Anniversary of DANIDA

Ravallion M, Chen S (2001) Measuring pro-poor growth. World Bank e library

Rostow WW (1960) Stages of economic growth. A non-communist manifesto. Cambridge University Press, New York, NY

Sachs I (2004) Inclusive development strategy in the era of globalization. ILO, Geneva

Sachs J (2015) The age of sustainable development. Columbia University. Earth Institute, New York, NY

Schumpeter J (1961) Theory of economic development (Translated from German). Oxford University Press, New York, NY

Sen A (1988) The concept of development. Chapter 1. Hand book of development economics. Elsevier, Amsterdam

Sen A (1999) Development as freedom, 1st edn. Oxford University Press, New York, NY

Steeten P (1979) Basic needs; Premises and promises. World Bank reprint series number 62. Washington, DC

Sutcliffe R (1980) Industry and underdevelopment. Andover, MA, Addison

The World Bank (2000) Attacking poverty World development report. The World Bank, Washington, DC

UNDP (1990) Human development report. Concepts and measurement. Oxford University Press and UNDP, New York, NY

UNDP (2014) Human development report. UNDP, New York, NY

UNDP (2000) Pro–poor growth; concepts and measurements with country case studies. New York, NY

UN (2013) Global sustainable development report. New York, NY

UN (2015) Economic report on Africa. United Nations Economic Commission for Africa, Addis Ababa

WCED (The World Commission on Environment and Development) (1987) Our common future. Oxford University Press, Oxford

WDI (2015) World Bank Development Indicators 2015. Washington, DC

White H, Anderson E (2001) Growth versus distribution. Does the pattern of growth matter? Dev Pol Rev 19:267–289

WSSC (World Social Science Report) (2013) International Social Science Council. WSSC, Paris

Part V
Conflict-Growth Relationship and Reconstruction

Benefit-Cost Analysis of the Integration of Rwanda in the East African Community: 2007–2013

Emmanuel Mushimiyimana

Abstract Regional integration is a policy that provides both political and economic gains for member states. However, to benefit from it requires that a member country uses integration opportunities while overcoming challenges. Knowing that Rwanda is a country torn by war and the genocide of 1994 perpetrated against Tutsi this paper finds that Rwanda is rebuilding itself through the East African Community (EAC) by applying political and economic reforms convenient for its integration. Through EAC Rwanda is ushering in both political and economic reforms though there are still challenges in both these fields. Politically, the will for a political federation is still limited. Economically, Rwanda is facing a trade deficit though there has been an increase in its trade relations with EAC in the last 5 years. This trade deficit is due to lack of enough goods for export in the EAC and also its low levels of production. However, due to economies of scale and competition, Rwanda is bringing in new products to the market and its production is improving domestically in terms of food processing and construction materials. Economic policies such as industrial creation in Rwanda and looking for alternatives to compensate for a decline in tariff revenue have also been necessary.

Keywords Rwanda • Regional integration • East African Community • Trade • Politics

1 Introduction

Regional integration provides both political and economic gains for member states. Political gains are security (Boutros-Ghali 1992: section VII) bargaining power (Grugel and Hout 1999), project cooperation, lock-in to reform (Acharya 1999) power and recognition. Economic opportunities are due to good effects of economies of scale, competition, trade and location. These are, for instance, (1) increase in production efficiency, (2) decrease in prices of goods and services,

E. Mushimiyimana (✉)
University of Rwanda, School of Social, Political and Admnistrative Sciences, Butare, Rwanda
e-mail: manemanu12@yahoo.fr

A. Heshmati (ed.), *Economic Integration, Currency Union, and Sustainable and Inclusive Growth in East Africa*, Advances in African Economic, Social and Political Development, DOI 10.1007/978-3-319-30432-8_11

and (3) increase of people's welfare (Fry and O' Hagan 2000: 117). However, these opportunities go hand in hand with challenges. The political challenges are: (1) possible divergence of national interest, (2) lack of sovereignty, and (3) lack of political will. The economic challenges include: (1) fall of custom tariff revenue especially when there is no compensation mechanism, and (2) trade deficit.

Previous research that I carried out when Rwanda was integrating in the EAC in 2007 highlighted the following results: (1) manufactured goods such as cosmetics and related products in the Rwandan market were mainly from Kenya. (2) Food products available in the Rwandan market were mainly from Uganda. In 2007, no cooking oil from Rwanda was available in the market, especially in Huye market (a focal area of that study). (3) Other consumable goods like choose, house materials and cement were from Uganda and Kenya. (4) Tanzania exported less to the Rwandan market. (5) Burundian products were almost not available in the Rwandan market. (6) Rwandan goods were so few and expensive that they could not even satisfy the domestic market (Mushimiyimana 2007).

Using both qualitative and quantitative methods and specifically basing the study on desk research, interviews and observations, I was able to respond to the following research questions:

1. What are the political and economic gains achieved during the 6 years of integration?
2. What are the remaining challenges that Rwanda is still facing in the community?
3. How can Rwanda overcome these challenges?

2 Literature Review

2.1 Political Opportunities and Related Challenges

2.1.1 Security Opportunities

The main political impetus for integrating has been security gains. The preamble to the 1951 treaty establishing the European Coal and Steel Community out of which the European Union grew, stated its aim as 'to create, by establishing an economic community, the basis for broader and deeper community among peoples long divided by bloody conflicts.' Before MERCOSUR, a Southern American regional integration comprising of Argentine, Brazil, Paraguay and Uruguay was created in 1991, Argentine and Brazil militaries had long perceived each other as potential threats. However, after economic agreements covering steel and automobiles were signed, tensions between the two reduced.

Security gains in both the European Union and MERCOSUR are explained by political scientists who suggest that negotiations between political leaders on trade issues gradually build trust. Therefore, elites learn to form cross-national coalitions for subsequent collaborations. Therefore, economic cooperation and interdependence between countries further decreases state-centered nationalism and hence leads to political integration. Besides, they may have precisely the opposite effect mainly when enhancing security is not the main concern or when trade becomes costly for one of the partner states. In that case tariff preferences that induce regional trade can create powerful income transfers within the region and can lead to the concentration of industries in a single location. The loss of income then creates resentful separatist movements which lead to conflict. This has been the case of the American civil war because of what the South called the tariff of abomination which caused a great transfer of incomes to the North to the detriment of the South.

The East African common market (now the East African Community) illustrates this conflict. Tanzania and Uganda complained about income transfers to Kenya created by the common external tariff on manufactures. There was agglomeration of the manufacturing industry in Kenya, compared to small industries in Tanzania and Uganda. Lack of compensation for income transfers contributed to the collapse of the common market, the closing of borders and the confiscation of community asserts in 1977–1978. That hostility contributed to the conflict between Tanzania and Uganda in 1978 (Schiff 2000) though there were other sources of it such as political misunderstandings between Idd Amin and Nyerere's regimes. Such a case has also been found in the Central American Common Market (Pomfret 1997).

From these examples, it is clear that regional integration sometimes promotes intra-regional security and sometimes worsens it. The European Union has avoided this problem. When a country signaled that a community policy would cause major problems for it, this was accommodated either by being compensated as the British budget relates, or by being granted a gradual adjustment process, as with compliance with the rules on labor mobility by Spain and Portugal (Tsoukalis 1993) or adopting the euro by England. No matter what the challenges, regional integration provides a real step towards intra-regional security.

States unite to face a common external threat. An example of this is the Gulf Cooperation Council (GCC) created in 1981 partly in response to the potential threat from regional powers such us Iran and Iraq (Kechichiam 1985). The Association of Southeast Asian Nations (ASEAN) was motivated by a perceived need to stem the threat of spreading communism in the region. Central and Eastern European countries joined EU for protection against the Russian Federation. In Africa, the Economic Community of West African States (ECOWAS) illustrates both internal and external security arrangements. In 1986, ECOWAS, with 16 members, ratified a mutual defence protocol that authorized military intervention by the community both in conflicts between members and if a conflict in a member country was instigated from outside the membership. The case of ECOWAS/ Economic Community of West African States Monitoring Group (ECOMOG) in the Liberian conflict serves as an example.

2.1.2 Bargaining Power

Power is an essential element in the politics of any country. Therefore, small countries can band together to increase their bargaining power in trade negotiations. The formation of the European Economic Community (EEC) in 1957 was aimed at increasing its bargaining power relative to the United States of America (Whaley 1996). It accelerated US-European trade liberalization in manufactures that gave access to American markets and so delayed trade liberalization in agriculture. This raised the incomes of European farmers.

Besides, developing country regional groupings have rarely been able to succeed in negotiating as a group in the league of high economic countries. This is due to their different politico-economic interests. Though individual countries may have so little bargaining power that they cannot negotiate attractive deals, regional groupings can achieve this. The Organization of Petroleum Exporting Countries (OPEC) has exploited this opportunity and been able to dominate oil supplies and fix petroleum prices according to real gains of purchase.

In fact, pooling forces and costs of negotiations should be crucial for the poor and small countries of Sub-Saharan Africa. Although they have little role in the international market or in global political influence, being united may attract negotiations or investors and donors. Rwanda within the framework of the EAC can negotiate with EU or the North American Free Trade Agreement (NAFTA) on trade but rarely can Rwanda herself attract some trade organizations to negotiate with.

2.1.3 Project Cooperation

Countries can benefit greatly from cooperation when they share resources such as rivers, lakes, fishing grounds, hydroelectric power and railway connections or when they join hands to overcome problems such as pollution and poor transportation. The South African Development Community (SADC) provided the Southern African Power Pool (SAPP) for regional electricity. EU and the East European and Mediterranean countries have numerous initiatives for cross-border projects covering transport networks, energy, environment and other infrastructure projects. EAC also has some projects such as Power Master Plan, railways and others that Rwanda as well as other member states can benefit from. These include electric dams, methane gas exploitation and fishing.

2.1.4 Lock-in to Reform

To have a membership in a certain regional integration helps a country reform. Integration itself can enforce or compel a country to reform. In this way, regional integration is used as a commitment mechanism. For instance, a non-democratic country cannot be admitted to the EU. Morocco has been denied entry because of this. Some Eastern Europe countries have democratized because of this incentive.

MERCOSUR fought against Paraguay's military coup in 1996. The bloc of four presidents backed by the US and the Organization of American States (OAS) reportedly quashed the rumored coup with a strong joint statement that democracy was a condition of membership to the bloc. Two months later MERCOSUR amended its charter to exclude members who had abandoned the full exercise of republican institutions (Presidential Declaration on the Democratic Commitment in MERCOSUR, Saint Louis, Argentina, June 25, 1996; Talbot 1996). Thereafter, Chile and Bolivia accepted democracy as a condition for membership. However, this condition is only binding if the penalties for violating it are severe and are effectively enforced or when integration is more important than exclusion for individual nation-states.

Reforms occur not only in politics but also in the economy where liberalization and budgetary and fiscal policies are renewed and amended. For instance, Mexico profited a lot in integrating in NAFTA where the US has been committed to supporting this country for reducing emigrants. Unfortunately, developing countries have failed to exploit this opportunity because of lack of credibility among them and lack of who was going to enforce it. South-South regional integration agreements are hardly democratic. Although developing countries have some form of elections, this does not mean that they are democracies.

2.2 Economic Opportunities with Regional Integration and Related Challenges

There are many economic opportunities underpinned by integration although challenges also exist. Both of these are seen in two economic mechanisms that accrue to integration: competition and scale, and trade and location. Competition and scale effects arise as separate national markets become more integrated in a single unified and open market. Producers in member countries enter into closer contact and competition with each other. The entrenched monopoly is then eroded; prices reduce and the flow of foreign direct investment (FDI) increases.

Besides, trade and location effects arise when a regional agreement changes the pattern of trade and the location of production. The direction of trade changes as imports from partner countries become cheaper, encouraging consumers to substitute these for imports from the rest of the world (trade diversion phenomena). Both of these effects create real income changes for consumers and producers, as well as in changing government tariff revenues. These changes bring about convergence (or divergence) of income levels.

Convergence occurs when income levels in poorer countries rise to the levels of the richer partners. On the contrary, divergence occurs when some countries gain at the expense of others. This challenge has been fought by using the asymmetric principal. For instance, in EAC, Uganda and Tanzania reduced tariffs to zero on customs for Kenyan export products in their markets after a period of 5 years, until 2009 whereas Kenya reduced customs on items from those two countries. The other mechanism used to overcome the divergence effect is compensation on government revenue deficit. This is done through regional integration institutions like regional banks or through mutual budgetary fiscal policies among member states. Unfortunately, most of the time this mechanism is ignored or badly executed in South-South regional integration and this hinders the progress of the integration.

2.2.1 Competition and Scale

Opportunities within this economic mechanism are many. First, because of increase in competition, firms are induced to cut prices and expand sales, benefiting consumers as monopolistic distortions are reduced. Second, the market becoming bigger allows firms to exploit economies of scale more fully. Third, competition and economies of scale induce firms to reduce internal inefficiencies.

Researchers like Chang and Winters (2002) indicate that increased competition in MERCOSUR markets induced exporters to cut prices, thereby improving the terms of trade of MERCOSUR countries and yielding a sizeable welfare gain. The internal reduction in prices due to free trade induces or compels a reduction in import prices from the rest of the world. Take for instance a Brazilian case. Prices were reduced even for the products that Argentine (Brazil's partner) does not have only for fear of competition. Terms of trade increased in MERCOSUR.

2.2.2 Regional Integration and Foreign Direct Investment

If regional integrations create large markets, they will attract FDI. Investors increase their activities as the market becomes more competitive, favoring lower marginal costs of production. They are also motivated by security, good governance and other values like stability and transparency. Mexico is an example of a country that benefited in terms of FDI because of integration in NAFTA. FDI generally increases economic activities in a country—it can raise revenues, employment and technology transfers. However, this happens when FDI is not enhanced for purposes of 'tariff jumping' or escaping taxes. Therefore, local production costs must not exceed the costs of imports. This means that local production costs should be low enough to make local products cheaper than imported ones.

2.2.3 Trade and Location

Imports from partner countries in regional integration become cheaper due to elimination of tariffs; demand patterns also change leading to changes in the flow of trade and in output levels. All these cause economic effects like contraction of some sectors or expansion of others and relocation of industries from one country to another. Preferential trade liberalization can induce trade creation or trade diversion. Trade creation occurs when a partner country's production cost is low so that products can be purchased within partner states without substituting them with products from the rest of the world, therefore, partners gain. On the contrary, trade diversion occurs when the partner country's production cost is high and partner states purchase goods from the rest of the world (Viner 1950).

Partner states within the region are advised to lower external tariffs as much as possible so that the cost of internal imports is cheaper unless imports from the rest of the world create trade diversions. Trade creation and trade diversion can also occur between countries of the same regional integration. People can buy cheaper products from one of the partner states and substitute them with their own expensive domestic products.

Imports and Exports with Trade Diversion

Suppose that a Rwandan imports a good from Kenya at 205 Rwandan francs and from China at 200 per unit, and the duty is 10 in both cases. When Rwanda integrates with Kenya the duty (10 Rwf) will be waived. Therefore, people will buy at 205 Rwf from Kenya whereas the Government of Rwanda loses 10 Rwf of tax revenue because of free trade. But the people (consumers) will gain 5 Rwf on the goods they would buy at 210 Rwf. The government will have a loss of 5 Rwf for a good it pays at 205 that it would otherwise pay 200. This is what we call the deleterious (harmful) welfare effect of 'trade diversion'. This example shows the impact of opening the market on citizens and on governmental revenue where people profit when there is a decrease in prices in regional economic community. However, governments are there for the good of their people.

Regional integration between developing countries is likely to only generate trade diversion and not trade creation. This is seen in homogeneous goods. In each member country, domestic consumer prices are fixed at the world prices plus import tariffs. Thus, each member country substitutes cheaper imports from the rest of the world to the detriment of expensive partner imports. The outcome is trade diversion and a loss for both countries (Schiff 1996). How will Rwanda manage the loss if trade deficit occurs in EAC? A country within a regional integration can try its best to lower the cost of exports, otherwise partners will lose incentives to purchase and then consumers will buy goods from the rest of the world.

2.2.4 Effects on Governmental Revenue

Both import and export trade diversions are challenges that affect countries that mostly depend on tariff revenues. In SADC, for instance, Zambia and Zimbabwe lost 5.6 and 9.8 % of government revenue respectively whereas South Africa gained. These are very substantial revenue losses. Therefore, countries are advised to first set up measures of alternative tax systems (like value added tax or VAT and compensation mechanisms) before removing sources of trade tax revenue from customs. Therefore, to integrate countries must look at how competitive they are and how to compensate the deficits that can occur. Rwanda is one of countries that still depends on customs tariff revenues. Actors in its integration policy must be aware of this and see how revenue losses do not affect the national budget through the compensation mechanism stated earlier.

2.2.5 Convergence or Divergence Due to Trade and Location

Integration can lead to relocation of economic activities; industries can agglomerate or expand in some countries or in others. This can affect a member state positively or negatively as labor or real incomes change. There is convergence in a regional integration when there are relative gains for all countries. There is divergence when there are some countries which gain at the expense of others due to relocation, agglomeration or trade diversions. Divergence is mainly seen in South-South Regional Integration. Divergence has also been found in EAC and in CEAO.

Uganda and Tanzania contend that in the 1970s all gains of the East African Common Market were going to Kenya, which was steadily enhancing its position as the industrial center of the common market, producing 70 % of the manufactures and exporting a growing percentage of them to its two relatively less industrial partners. By 1958, 404 of the 474 companies registered in East Africa were located in Kenya. The community collapsed in 1977 because it failed to satisfy the poorer members which were getting an unfair share of the gains. Kenya gained because of it comparative advantage in manufacturing, relocations and access to relative free markets in Uganda and Tanzania. So Uganda and Tanzania suffered from divergence. These phenomena can be caused either by advanced technology, geographical gains, infrastructural facilities and endowments of capital and labor that are unequal within the region. Agglomeration is then clustering of economic activities in one area. It arises from an interaction between centripetal forces, encouraging firms to locate close to each other and centrifugal forces, encouraging them to spread out.

According to Marshall (1920), centripetal forces are classified in three groups. The first is knowledge pullovers, or other beneficial technological externalities that make attractive firms locate close to each other. The second is labor market pooling and the third are linkages between buyers and sellers where firms locate closely to their suppliers and buyers (consumers). On the other hand, centrifugal forces are due to congestion, pollution or other negative externalities associated with concentration of firms' economic activities.

A country can benefit from regional integration simply because of its production network backed by knowledge flows. Evidence shows that Ireland has attracted multinationals, which use the country as an export platform to supply to the rest of Europe. By 1993, foreign plants were producing 60 % of gross output and accounted for nearly 45 % of manufacturing employment (Barry and Bradley 1997). They were concentrated in the high-technology sector, importing two-thirds as inputs and exporting 86 % as output. By improving its science and technology and the knowledge-based economy, Rwanda too can attract investors. Nevertheless, advanced countries like Kenya and Uganda can occupy the production market in EAC so that Rwanda loses. Therefore, care must be taken to avoid falling in divergence.

2.3 About Policy Choices

Regional integration policy choices fall under four headings. The first is with whom (single or many partners). The second are the external policies of countries in a regional integration, whether it is a free trade area or a customs union. The third is depth of integration and fourth is width of integration. Researchers find that forming a regional integration with many countries with different regional policies can bring complexities or contradictions of policies and expensive contributions. Therefore, these challenges must be looked into before signing integration agreements and the government must choose the very profitable blocs.

Free trade is a good policy that not only inhibits wars due to protectionism, mercantilism (a case of the opium war) but also increases social welfare. Further, a customs union brings more incentives to real political integration, but there are many costs of being in a customs union. First, harmonization of external trade policies means a loss of national autonomy or sovereignty. Second, there can be a political divisive redistribution of revenues due to a common external tariff and lack of consensual ways of resolving disagreements. In developing regional integration, a customs union's difficulties include fixing a common external tariff and redistribution of revenues (case of CARICOM and CACM).

In short, it is important to understand regional integration, categorizing it and knowing what opportunities and challenges, both political and economic, countries face in a regional integration and vice versa. Since we know about the policies in regionalism and can balance them for the interests of nations and regional integrations I now discuss in practical terms what Rwanda is gaining in EAC at what cost and how policies of integration can be managed considering the challenges that the country is facing.

3 About the East African Community

EAC is a regional intergovernmental organization of the Republic of Kenya, Tanzania, Uganda, Rwanda and Burundi; the last two countries joined EAC recently. EAC has about 1.9 million square kilometers, a population of around 115 million people and a combined gross domestic product of more than US$41 billion. In financial year 2007–2008, the budget was US$28,313,559. This regional community has a vision, objectives, structures, principles and strategies for its economic and political integration which it hopes to realize through its organs and institutions in Arusha.

3.1 Institutions

The treaty, under Article 9 provides for the creation of the following organs and institutions of the community: (1) Summit, (2) Council, (3) coordination committees, (4) sectional committees, (5) the East African Court of Justice (EACJ), (6) the East African Legislative Assembly (EALA), and (7) Secretariat, and such other organs and institutions as may be established by the summit. In this regard, the following have been established: (1) the East African Development Bank, (2) Inter-University Council of East Africa, (3) Lake Victoria Basin organization, and (4) East African Centers of Excellence (Dar-es-salaam Maritime Institute, Utali College, Saroti Flying Academy). The fifth extraordinary summit of EAC established the following commissions: (1) The East African Science and Technology Commission, (2) The East African Civil Aviation Safety and Security Oversight Agency, (3) The East African Kiswahili Commission, and (4) The East African Health and Research Commission.

3.2 Fundamental Principles of the Community

Article 6 of the Charter has the following fundamental principles: (1) Mutual trust, political will and sovereign equality, (2) Peaceful coexistence and good neighborliness, (3) Peaceful settlement of disputes, (4) Good governance including adherence to the principles of democracy, the rule of law, accountability, transparency, social justice, equal opportunities, gender equality, as well as the recognition, promotion and protection of human and people's rights in accordance with the provisions of the African Charter on Human and People' s Rights, (5) Equitable distribution of benefits, and (6) Cooperation for mutual benefit.

3.3 Operational Principles of EAC

Article 7 provides the principles that govern the practical achievement of the objectives of the community. These include: (1) People centered and market-driven cooperation, (2) Provisions by partner states of an adequate and appropriate enabling environment such as conducive policies and basic infrastructure, (3) Establishing an export oriented economy for partner states in which there shall be the movement of goods, persons, labor, services, capital, information and technology, (4) The principle of subsidiarity with an emphasis on multi-level participation and the involvement of a wide range of state-holders in the process of integration, (5) The principle of variable geometry which allows for progression in cooperation among groups within the community for wider integration schemes in various fields and at different speeds; (6) Equitable distribution of benefits accruing to or to be delivered from the operations of the community and measures to address economic imbalances that may arise from such operations, (7) The principle of complementarities, and (8) The principle of asymmetry.

3.3.1 Objectives of the Community

Article 5 sets out the following EAC objectives: (1) Attaining sustainable growth and development of partner states by promoting their more balanced and harmonious development, (2) Strengthening and consolidation of cooperation in agreed fields that will lead to equitable economic development within the partner states and which will in turn raise the standard of living and improve the quality of life of their populations, (3) Promoting sustainable utilization of natural resources of partner states and taking up measures that will effectively protect the natural environment of the partner states, (4) Strengthening and consolidating long standing political, economic, social, cultural and traditional ties and associations between the peoples of partner states so as to promote people-centered mutual development of these ties and associations, (5) Mainstreaming of gender in all its endeavors and enhancing the role of women in culture, social, political, economic and technological development, (6) Promoting peace, security and stability and good neighborliness among the partner states, (7) Enhancing and strengthening partnerships with the private sector and civil society in order to achieve sustainable socioeconomic and political development, and (8) Undertaking such other activities of the communities as the partner states may from time to time decide to undertake in common.

3.4 Sources of Funds for EAC

The main source of funds for EAC is equal contributions by partner states and development partners. However, Article 133 allows the community to raise funds from other sources such as grants, donations, project funds and incomes. In April 2006 a group of experts recommended other possible sources of funds such as: (1) Giving EAC the mandate over specific areas in revenue collection related to EAC services, for example, processing East African passports and other immigration services, (2) Revenue derived from forestry and timber products in ecologically protected zones and tourism, (3) Import duties—1 % payable on luxury items, for example, luxury cars, sports equipment and jewelry, (4) Export duties, for example, on fish exports because inputs in terms of production are negligible at 0.25 %,

(5) Levies on some service industries such as airlines, fuel, cable television (0.5 % of VAT contributions), (6) Principle of one time investment endowment raised from bilateral and multilateral organizations such as the World Bank, (7) Donations, (8) Partner states' contributions based on 1 % gross national product (GNP), and (9) Rent from EAC's fixed assets.

3.5 External Trade in the Region

External trade exports of EAC are mainly agricultural products such as horticulture, tea, coffee, cotton, tobacco, pyrethrum, fish, hides and skins. Other types of exports include handicrafts and minerals such as gold, diamonds, gemstones, soda ash and limestone. Tourism is another economic activity that EAC exchanges with foreign countries. Imports are mainly from the European Union, Japan, China, India, the United Arab Emirates (UAE) and Saudi Arabia. These imports are mainly machinery and capital equipment, industrial supplies and raw materials, motor vehicles and their parts, fertilizers and crude and refined petroleum products.

It is clear that since 2002, imports from SADC have exceeded EAC exports in that regional community. This has led to a deficit in EAC's balance of trade to SADC's benefit. However, EAC has been able to garner more exports than imports in COMESA and this has accounted for a benefit in the balance of trade since 2002 (EAC Secretariat, Sectoral Council on Trade, Finance and Investment Paper, June 2007: 11).

3.6 Areas of Cooperation

The areas of cooperation are broad; they cover trade, investments and industrial development, monetary affairs, human resources, science and technology, agriculture and food security, environment and natural resources management, tourism and wildlife management, health, social and cultural activities and the free movement of factors of production. Politically, cooperation goes further in defense, juridical affairs and leadership (in democracy, human rights, gender and political stability).

3.7 Partnerships

In the global arena, EAC collaborates with other African organizations in the spirit of the Abuja treaty and in accordance with the second conference of African Union Ministers in charge of Integration on Rationalization of Regional Economic Communities in order to prepare for AU Economic and Political Integration. With financial partners, EAC cooperates with the Africa Development Bank, IMF and the World Bank, and adheres to WTO arrangements. It cooperates with the UN, USA, European Union and other states and international organizations worldwide. It has partnerships with AU, COMESA, Inter-governmental Authority on Development (IGAD) and the Southern African Development Community (SADC).

3.8 EAC's Perspective

EAC as any other regional community has stages of integration. Generally, there are five stages of integration. The first is a 'Preferential Trade Area' where countries maintain their own external tariffs on goods imported from third countries. The second is a 'Free Trade Area' where they set up the rules of origin; the principle states that goods from partner states should be pay-free but their supplementary value from the rest of world should be paid for. Then, states begin to harmonize their external tariffs. In EAC, a good is considered to be from a partner state if at least 35 % of its inputs are from that country (interview, RRA, Customs Operation, 2007). The third is a 'Customs Union' that allows member states to adopt a common tariff nomenclature (CTN) and operate a common external tariff (CET) and adopt common customs legislations. Regulations and procedures include those relating to the value of imports, documentation and clearances required for imports. A customs union provides that all imports are subject to a common community policy. Then, common community decisions regulate tariff rates, exemptions and key customs procedures.

Fourth is a 'Common Market' stage where free movement of people, goods and services, labor and capital and the right of establishment and residency between member states of the common market are accepted. The last stage is 'Political Federation' or the formation of only one federal state, politically controlled by one government with its federated states.

EAC adopted the customs union in 2005. On 20 November 2009, EAC member states signed the common market protocol that includes free movement of people and establishing a single currency. Its Article 104 stipulates establishing free movement of persons, labor services and right of establishment and residence in EAC. In July 2010, member states ratified that treaty and so they have an obligation to change their national policies to comply with what they have signed for.

The mechanism established for enabling free movement of people was the introduction of 'third generation' identity cards. These aimed at showing dual citizenship: one of the home country and the other of 'East Africa'. There should be one border post across the five partner states. The other regional integration that has progressed at this level of free movement of persons across its boundaries is the European Union. For instance, to cross from France to Belgium or from Belgium to Netherlands you do not need to have another visa. However, some countries like the United Kingdom in EU have set up protectionist mechanisms to limit free movement. In EAC, all countries did not accept fast-tracking the free movement of people using IDs as travel documents. Only Kenya, Uganda and Rwanda agreed to introduce this system. Tanzania and Burundi were reluctant to do so.

In January 2014, the citizens of Rwanda, Uganda and Kenya began using identity cards to travel in these three countries. Ugandans who did not have IDs, were allowed to use voter cards as valid travel documents while students were allowed to use student cards. This conclusion was taken from the EAC ministers responsible for the integration of Uganda, Kenya and Rwanda in August 2013. The meeting also proposed establishing a single tourist visa (MINEAC 2013).

4 Methodology

This research is both qualitative and quantitative (Q^2). The data given in this paper is mainly qualitative. I went to the market to find out about the availability of goods, their origins and their prices. I also labeled them into three categories: manufactured cosmetics (toothpaste, glycerin, soaps and other house materials), food (maize and wheat flour, rice and cooking oil as well as drinks), and other commodities such as construction materials (cement, stir reinforcement bars, etc.).

They were qualitatively interpreted, for instance, as 'many', 'few' or 'not available' in the lens of availability of products in the market. In this paper I did not focus attention on the affordability of goods and services like in my previous research in 2007 since the availability of goods and services determine their prices, bargains and choice of customers. When the commodities are not available they are, of course, expensive or do not exist in a given market study area.

The research is also quantitative as highlighted in Table 1. Quantitative data provided by the National Bank of Rwanda and the Rwanda Revenue Authority were gathered and interpreted. Using 'Working with Data' techniques the current shape of Rwanda's trade relations with the rest of the East African member states were also arrived at and interpreted. Interviews were used to supplement the observations, following a constructivist approach. A desk research was also done and the Ministry of Foreign Affairs and Cooperation of Rwanda was visited. Focused interviews with leaders and businessmen in Rwanda, especially in the trade, fiscal, security and governance sectors were also done. Data on commodities were gathered through observations of the Huye market as a sample. Huye is one of the bigger markets in Rwanda located in the southern province. It started in 1924 and was first called Astrida market and then Butare market.

The research is evaluative and tries to determine EAC's role in the reconstruction of Rwanda. I had done similar research in 2007 when my analysis focused on what Rwanda would gain if it became a member of the EAC. Rwanda and Burundi became members of EAC in 2007. The current research was done 6 years after Rwanda became a member of EAC to find out what it has already gained and what it has lost. The limits of this research are that some data of it in the economy is

Table 1 Comparison of trading and strategic distances in the region

Distance from	Distance (km)	Implication/interpretation of political, social and economic management in the EAC region
Eastern DRC an insecure region for EAC international trading/transportation ports		
Goma/Bukavu to Mombasa (Kenya)	1187/ 1211	In terms of trade and transport, east DRC is closer to the eastern ports of EAC and depends more on Rwanda, Burundi, Kenya and Tanzania than on western DRC. Insecurity in those countries may severely affect livelihoods and development of east Congolese citizens as well as of those living in EAC.
Goma/Bukavu to Dar Es Salam (Tanzania)	1250/ 1242	
Goma/Bukavu to Matadi (DRC)	1806/ 1744	
Eastern DRC to capital/administrative cities in EAC		
Goma/Bukavu to Kigali (Rwanda-EAC)	96.7/ 145.8	Eastern DRC is strategically closer to Kigali and Bujumbura (EAC) than it is to its central capital Kinshasa. This implies that security in eastern DRC is, or should be, a threat and concern for Burundi and Rwanda (EAC). If Rwanda and Burundi are threatened all trading and other strategic interests of Kenya, Uganda and Tanzania are automatically hampered.
Bukavu to Bujumbura Rural Province (Burundi-EAC)	115.8	
Goma/Bukavu (Eastern DRC) to Kinshasa (DRC capital city)	1575/ 1519	

Source: Mushimiyimana (2011), measures of distances are from True Knowledge

quantifiable while some data like that in politics is not. Triangulation is used to increase the validity of the data by basing it on ground realities.

5 Political and Economic Gains Achieved: 2007–2013

5.1 Political Gains

5.1.1 Security

Security in Rwanda is of concern for all EAC member states and partners due to her strategic location in the region. Insecurity in Rwanda directly affects Kenya, Uganda and Tanzania, Burundi and the eastern part of the Democratic Republic of Congo (DRC): North and South Kivu. Table 1 highlights Rwanda's strategic location which guarantees that a collective defense policy is a factor for EAC though collectively enforcing security is still a low priority as compared to the other regions. The regional reference is to Eastern DRC towns not only because they are trading partners with EAC, or are joined to EAC by Rwanda and Burundi, but also because they are regionally insecure and that their insecurity can spill over to EAC.

The case of EAC illustrates that security in the region must be seen as a serious issue to counter-balance regional threats such as terrorism and rebellions as well as other wars related to the region's 'resource curse' (existence of minerals and oil) and conflict trap (existence of cross-border ethnic violence). EAC's geopolitics and geostrategic situation as a neighbor to the horn of Africa (Somalia) and the regions in which we still find structural ethnic violence (Sudan, DRC, etc.) dictate common decision making and attract collective defense concerns.

Rwanda has benefited much in terms of both intra-regional and extra-regional security in the EAC. Since its integration, the people of Rwanda have a sense of another identity of being East African; Kiswahili is taught in schools and a movement of people to Uganda, Kenya and Tanzania is increasing cross-cutting cleavages. Cross-cutting cleavages are important not only for Rwanda but also for Burundi. The bipolarization of ethnicity (Hutu and Tutsi identities) has increased domestic violence in recent times in these countries. Rwanda is combating a former constructed ethnic identity; however, the success of this will require a lot of effort at both integration and domestic open competition between the Rwandans and other East African citizens. What is meant by more integration is free movement of people, goods and services and the creation of a EAC common identity and citizenship.

5.2 Bargaining Power

EAC has provided to Rwanda potential intra and extra-regional bargaining power. Through the European Partnership Agreement (EPA), EAC will export to the European Union free of charge on tariff duties with an asymmetric exchange whereas EU commodities will gradually pay tariff to enter into the EAC market. It is for Rwandans to reap the benefits that this opportunity provides by providing goods for export. There is also the Trade and Investment Framework Agreement with the US through EAC:

> (T)he EAC signed a regional 'interim' agreement with the EU at the end of 2007. The 'interim' EPA gives the EAC members duty-free and quota-free access to the EU market in exchange for an asymmetric and gradual opening of their own economies to EU imports. The EAC continues to negotiate a comprehensive EPA with the EU. The EAC also signed a Trade and Investment Framework (TIFA) with the United States in July 2008 to strengthen economic relations between the two parties (the World Bank 2010).

In terms of migration and remittances, EAC hosts a large Rwandan diaspora thereby providing a real potential area for intra-regional bargaining (Table 2). This provides the option of repatriation for those who still have refugee status or those who need other civil status through the cessation clause. Rwanda also receives remittances from this diaspora.

Table 2 shows that most Rwandan migrants are in the EAC community and this has both political and economic implications—the importance of EAC for Rwanda and the voice of the latter in intra and extra-regional politics and the economy.

Table 2 More than 70 % of Rwandan Diasporas live in EAC

EAC countries	Burundi	3117	183,195	76.1 %
	Kenya	11,750		
	Tanzania	67,071		
	Uganda	101,257		
Other world regions	Belgium	6792	57,566	23.9 %
	Germany	4496		
	Ethiopia	4261		
	DR Congo	3503		
	USA	2800		
	France	2721		
	Pakistan	2681		
	Canada	2680		
	UK	2632		
	Others	25,000		
Total		240,761	240,761	100 %

Source: Development Research Centre on Migration, Globalisation and Poverty (March 2007) and Orozco (2009)

5.3 *Project Cooperation*

Countries can benefit greatly from cooperation when they share resources like rivers, lakes, fishing ponds, hydro-electric power and railway connections; or when they work hand in hand to overcome problems including pollution and poor transportation. European Union and the East European and Mediterranean countries have numerous initiatives for cross-boundary projects covering transport networks, energy, the environment and other infrastructure projects. EAC also has some projects such as the Power Master Plan and railways that Rwanda as well as other member states can benefit from. These also include electric dams, ICT infrastructure, methane gas, oil exploitation and fishing. Unfortunately, this mutual exploitation of resources and joint infrastructural build-up is still limited and does not have an impact on the Rwandan economy. Both political will and financial capacity are needed for joint implementation and for raising funds for such huge projects, either through international financial agencies or through development oriented donors.

5.4 *Lock-in to Reforms*

Being a member of a certain regional integration can help a country undertake reforms. Integration itself can enforce or compel a country to reform. In this way, regional integration is used as a commitment mechanism; and it is in this regard that Rwanda has been able to reform its education system by introducing Kiswahili and

English in teaching. Rwanda is also adopting stability and governance reforms that lead to inclusive development and avoidance of ethnic and regional discrimination through the EAC experience. These are core values of EAC that have been helping its member states to overcome domestic violence. Furthermore, Rwanda as well as other EAC member have also tried to counter-balance the escalation in ethnic violence that followed the 2008 elections in Kenya.

As for the economy, Rwanda has adopted innovative behavior since 2007, due to direct and open competition; services and goods in Rwanda are improving and have at least reached the same level as Ugandan and Kenyan products. The establishment of banks such as KCB and other services like education (Ugandan and Kenyan universities in Rwanda) have led to crucial reforms in both banking and academic sectors that directly impact people's welfare and development as well.

5.5 Competition and Economies of Scale

Opportunities within this economic mechanism are many. First, thanks to an increase in competition, firms have been induced to cut prices and expand sales, benefiting consumers as monopolistic distortions have reduced. Second, an increase in the market allows firms to exploit economies of scale more fully. Third, competition and economies of scale induce firms to reduce internal inefficiencies.

Though the Rwandan economy is still strong as compared to the Kenyan, Ugandan and Tanzanian economies its local market is improving due to competition. During my interviews with businessmen in the Huye local market they said that since 2007 Rwanda had relatively increased its production and innovative marketing and selling. Some new products have also been put on the market like those from food processing (maize flour, rice) and house construction materials (such as steel reinforcement bars).

From Table 3 it can be seen that (1) manufactured goods such as cosmetics and related products in the Rwandan market are mainly from Kenya. (2) Food products available in the Rwandan market are mainly from Uganda. (3) Very few consumer goods like shoes, house materials, supplement cements are from Uganda and Kenya. (4) Tanzania exports less to the Rwandan market. (5) Burundian products

Table 3 Availability of goods from EAC in Rwandan market (2012)

	Made in				
Goods	Burundi	Kenya	Rwanda	Tanzania	Uganda
Cosmetics, toothpastes, soaps, glycerin, body lotions	Not available	Many	Few	Few	Many
Food (cooking oil, rice, maize and wheat flour)	Not available	Many	Few	Few	Many
Other products (house materials, shoes, cement)	Not available	Many	Few	Very few	Many

Source: Primary data collected by the author

are not available in the Rwandan market. (6) Rwandan goods are so few that they do not even satisfy domestic consumption.

Despite 6 years of integration, the situation in the Rwandan domestic market is quite similar to what it was at the beginning of 2007. There have only been slight changes. Kenya and Uganda hold a big share of EAC goods in the local market in Rwanda (estimated to be more than 80 %). This shows that Rwanda is experiencing a deficit in trade relations with EAC; this is also highlighted in the EAC trade report of 2008–2009 (EAC 2010) and this deficit is due to the industrial development gap in Rwanda as compared to Uganda in the food industry and as compared to Kenya in the manufacturing industry.

If Rwandan producers are not able to satisfy the domestic market, they will not be able to export on EAC's scale. Rwandan products are still expensive and scarce and so cannot be productive and competitive in the region. Quantitative data from the Rwanda National Bank also supports this argument.

Imports from EAC have increased considerably while exports have increased marginally and that too at a very slow pace (Table 4). This reflects the on-going deficit in trade relations between Rwanda and EAC. Rwanda needs to increase its production levels and define the areas of specialization more clearly. Table 4 also shows that almost all areas of production of goods are lagging behind. Small and large scale industries are needed for the Rwandan economy to flourish. Easy and cheaper transport facilities like railways should be a priority for the government and for other stakeholders like the private sector and donors.

This research found that: (1) Rwanda has trade deficit with partner states (mainly Kenya and Uganda). (2) Producers in Rwanda are improving their production standards but this improvement is relatively slow and cannot cover the losses. (3) Rwandans are profiting at least when it comes to welfare opportunities because of Rwandan integration with EAC as they are buying products from Kenya and Uganda more than before after the country opened its borders by signing the EAC Common External Tariff in 2009. EAC's Common External Tariff (CET) has a double effect; one positive and another negative. It is positive since it has reaffirmed Ricardo's theory of welfare benefits of an open market for people who access goods and service at cheaper prices; it has a negative effect because the Rwandan government has lost 25 % in tariff revenue since the implementation of EAC CET (the World Bank 2010).

Foreign direct investment has increased tremendously in Rwanda since 2007. This is due to internal economic policies and political stability, rather than because

Table 4 Evolution of Rwanda's trade with EAC (2006–2010) in USD million

	2006	2007	2008	2009	2010
Exports from Rwanda to EAC	36.51	45.03	46.25	47.34	54.16
Imports to Rwanda from EAC	241.73	316.17	461.1	449.65	513.35
Trade balance	−205.22	−271.14	−414.85	−402.31	−459.19
Total trade with EAC	278.24	361.2	507.35	496.99	567.51

Source: BNR (2011), Statistics department, MINICOM

of its integration with EAC, because Burundi integrated in 2007 as did Rwanda, but the former has not had such a positive increase in FDI inflows. In 2009, Rwanda's business environment improved significantly and its Ease of Doing Business rank for 2010 went to 67th out of 183 countries from 143rd in 2009 (the World Bank 2010: 4). Regional economic reforms of course must go hand in hand with domestic reforms to be more effective and 'Rwanda is now the top global reformer and for the first time for an African country' (Rusuhuzwa and Baricako 2009).

6 Policy Recommendations

A number of suggestions emerged from my interviews. Retailers that I spoke to told me that they were competing with some industries that were also retailing. They argued that this was not the case in Uganda or Kenya as the industry network worked with retailers to access the market instead of competing with retailers. They also added that there are still very few industries in Rwanda which can satisfy even the local market though there have been noticeable innovations since 2007. According to them tax is high domestically and this does help in smoothening business at the local level. Further, there is no large market in Rwanda where people from different countries can come and buy commodities. For instance, businessmen from Congo, Burundi and Rwanda meet in Uganda instead. There is a need to empower regional trading systems especially in Rwanda; this will be supported by the COMESSA trading hub which is coming up in Kigali.

Increasing internal taxes is an alternative way of compensating the decrease in custom tariff revenue due to EAC CET. However, this will impact businessmen a lot who in any case feel the burden of taxes. It will also impact people who do not fully enjoy a decrease in prices due to internal taxes and cost of transport. There is also lack of sufficient internal products which can substitute and compete with products from Kenya and Uganda. It is as if the two countries still enjoy a monopoly over the Rwandan market and this creates a monopoly in an open and free regional market system. Therefore, Rwanda does not need to focus on domestic tax revenue but on creating an easy business environment and setting up small and large scale industries domestically to increase the production of goods for export in EAC. This will also trigger domestic competition which will decrease the prices of goods and services.

The use of asymmetric principles and rules of origin are currently not sufficient to counterbalance the trade diversion between Rwanda and EAC. The Government of Rwanda must help the private sector to improve production and introduce public-private enterprises in areas where the private sector has failed.

EAC needs to work for joint implementation of regional infrastructure so that it has a greater impact on the economy and the welfare of its citizens. International financial agencies and donors may help implement projects like oil pipelines and railways to cut transport costs and hence decrease the prices of goods.

7 Conclusion

The EAC is strategically and naturally suitable for Rwanda's efficient and effective integration policy. Rwanda and Burundi's security are a matter of concern for all EAC member states as a threat to Rwanda and Burundi directly affects other EAC members especially in trade with other Central African states.

Rwanda has been historically torn by war and genocide due to protracted ethnic conflict and bad governance but now it is building a new identity for Rwandans: multi-cultural citizens economically oriented who are open to trade, business and education. This cross-cutting cleavages effect will be effective as further integration happens and as the will to integrate Rwandans with EAC persists.

Foreign direct investment and trade between Rwanda and EAC are increasing and there has also been an increase in innovations in the production of goods and services in Rwanda, especially in food processing and house construction materials. However, Rwandan goods and services are still scarce and expensive so as to be able to compete in EAC. This creates a trade deficit for Rwanda in EAC though generally Rwanda's trading relations with EAC have been increasing.

Since the year of its integration (2007), Rwanda has been gaining political and economic opportunities in EAC. Rwanda also needs to continue domestic economic reforms and efficiency to boost domestic production and decrease its trade deficit with EAC.

Researchers need to analyse deeply areas of specialization that Rwanda needs to have as an EAC member. If it is tourism, then how can people from EAC be motivated to come to Rwanda as tourists? If it is provision of services then how can EAC people be attracted with banking, insurance or ICT services in Rwanda? How can Rwandans reach the EAC labor market like other partner countries? If Rwanda needs to be a hub of EAC trade, then how does it move to achieve this? How can she create a good environment and be a gateway to EAC for Central Africa, West Africa and the rest of the world? Geographically Rwanda has a chance of improving the mining sector. Research and debate are open on this. Kenya is ahead in manufacturing and Uganda in the food and agricultural industry. Therefore, if Rwanda invests in mining this will be a good combination.

References

Acharya A (1999) After the global crises: what next for regionalism? University of Warwick, Coventry

Barry F, Bradley J (1997) FDI and trade: the Irish host-country experience. Econ J 107 (445):1798–1811

BNR (2011) Annual report, 2010, 2011. http://www.bnr.rw/publications.aspx

Boutros-Ghali B (1992) Agenda for peace: preventive diplomacy, peacemaking, and peace keeping. United Nations, New York, NY

Chang W, Winters LA (2002) How regional blocs affects excluded countries: the price effects of MERCOSUR. Am Econ Rev 92(4):889–904

Development Research Centre on Migration, Globalisation and Poverty (2007) ‘Global Migrant Origin Database’, Version iv. http://www.migrationdrc.org/
EAC (2010) East Africa trade report 2008. EAC Secretariat, Arusha, Tanzania
Fry G, O’ Hagan J (2000) Contending images of world politics. St Martin’s Press, LLC, New York, NY
Grugel J, Hout W (1999) Regionalism across the north-south divide: state strategies in the semi-periphery. Routledge, London
Kechichiam J (1985) The gulf cooperation council: the search for security. Third World Q 7 (4):853–881
Marshall A (1920) Principles of economics. MacMillan, London
MINEAC (2013) Tripartite ministers endorse single tourist visa by identity card. http://www.mineac.gov.rw/index.php?id
Mushimiyimana E (2007) Opportunities and challenges with regional integration: what does Rwanda hope to gain from being a member of the East African Community? Memoir. National University of Rwanda, Main Library, Butare, Rwanda
Mushimiyimana E (2011) Divergent interests of actors in the Great lakes region: a threat to peace and development in the east DRC and in the whole region. MSc Dissertation, Institute of Development Policy and Management, Antwerp, Belgium
Orozco M (2009) Emerging markets for Rwanda: remittance transfers, its market place and financial intermediation. Report, Inter-American Dialogue, Washington, DC
Pomfret R (1997) The economics of regional trading arrangements. Oxford University Press, New York, NY
Rusuhuzwa K, Baricako J (2009) The global financial crisis, slowing private capital inflows and economic growth in Rwanda and Burundi. African Economic Conference, Addis Ababa, 11–13 Nov 2009
Schiff M (1996) Small is beautiful preferential trade agreements and the impact of country size, market share, and smuggling. J Econ Integrat 12:359–387
Schiff M (2000) Multilateral trade liberalization, political disintegration, and the choice of free trade areas versus customs unions. Policy Research Working Paper Series no. 2501. The World Bank, Washington, DC
Talbot S (1996) Democracy and the national interest. Foreign Aff 75:47–63
The World Bank (2010) Rwanda trade in brief world trade indicators 2009/2010. http://info.worldbank.org/etools/wti/docs/Rwanda_brief.pdf
Tsoukalis L (1993) The new European economy: the politics and economics of European integration. Oxford University Press, Oxford
Viner J (1950) The customs unions issue. Carnegie Endowment for International Peace, Washington, DC
Whaley J (1996) Why do countries seek regional trade agreements? In: Frankel J (ed) The regionalization of the world economy. Chicago University Press, Chicago, pp 63–90

The Conflict-Growth Nexus in Sub-Saharan Africa

Syed Mansoob Murshed

Abstract Conflict in Africa has been explained by factors related to greed and grievances. These are insufficient to initiate conflict in the absence of institutional failure or a degenerating social contract which may be heightened by the lack of economic growth. I emphasize the *inseparability* between economics and politics, drawing out the similarities between the causes of conflict and the reasons for the lack of sustained growth both of which require institutional malfunctioning. The centrality of reducing inequalities, particularly categorical inequalities between groups based on unequal access to productive assets such as land, education, as well as individual inequality of opportunity, cannot be overemphasized. The democratic transition has the potential of producing violence as people have greater scope for venting dissatisfaction, especially when unaccompanied by egalitarian and pro-poor economic progress. The relationship between growth and conflict is non-linear; lack of growth and the poverty it produces engenders conflict. Equally rapid growth accompanied by heightening inequality can also cause conflict.

Keywords Causes of conflict • Causes of growth • Conflict-growth nexus • Institutions • Inequality of opportunity • Categorical inequality

---

This paper draws on my unpublished background paper prepared for African Development Bank's annual report (Murshed 2008). I wish to thank Muhammad Badiuzzaman for excellent research assistance.

S.M. Murshed (✉)
International Institute of Social Studies (ISS) at the Erasmus University Rotterdam (EUR), PO Box 29776, 2502 LT The Hague, The Netherlands

Department of Economics, Finance and Accountancy, Coventry University, Priory Street, Coventry CV 1 5 PB, UK
e-mail: Murshed@iss.nl

A. Heshmati (ed.), *Economic Integration, Currency Union, and Sustainable and Inclusive Growth in East Africa*, Advances in African Economic, Social and Political Development, DOI 10.1007/978-3-319-30432-8_12

1 Introduction

In the last two decades of the twentieth century Africa experienced a growth and development tragedy. This was accompanied by widespread civil war and other forms of large-scale internal conflict that accounted for over 1.6 million battle related deaths alone (PRIO-Uppsala dataset). Myrdal (1968) has written about the difficulties of accelerating growth in Asian countries such as India and Indonesia, and the stupendous effort needed to combat the sheer magnitude of endemic poverty in that populous continent[1]. In a similar vein, today's scholars and savants are busily engaged in addressing Africa's contemporary problems (Collier 2007; Sachs 2005, for example). Indonesia and other parts of East Asia also experienced large scale conflict during the 1960s. Since then, East Asia has sped ahead in terms of growth and poverty reduction, and several of its conflicts have been resolved or contained. Since the beginning of the new millennium, Africa has reversed the trend of 1980–2000: growth has picked up and substantial conflict abatement has taken place. Are both poverty and conflict symptoms of a deeper malaise, the lack of economic progress? Does peace building in contemporary Africa involve solutions to the problems of inequality and poverty that require us to delve deeper into the determinants of long-run growth? I argue that this is, indeed, the case.

Fundamentally, violent internal conflict implies the breakdown of institutional mechanisms for the peaceful resolution of conflict, which is one of the chief requirements for the mobilization of antagonistic groups for conflict. I have referred to these mechanisms for dispute resolution as implicit or explicit 'social' contracts (Murshed 2010). Civil war creates humanitarian crises that are of concern to the international community; they also contribute to global and regional insecurity. Economic policies cannot be formed independently of their consequences for social conflict. Similarly, the potential for conflict and civil war retarding growth and development is equally important. Thus, there is an inherent inseparability between economics and politics.

Conflict does retard the pace of economic progress. Murdoch and Sandler (2004) assess the adverse impact of civil war at home as well as in neighboring countries on a country's growth performance. They find that a civil war at home can reduce a country's growth by 31 % in the long run and by 85 % in the short run. Moreover, each additional neighboring civil war can lower growth by approximately 30 % of the host country effect in the long run and by 24 % of the host country effect in the short run. There is an even greater consensus about the converse; the risk of civil war is heightened by low growth and endemic poverty (see, for example, Collier et al. 2003).

But where does the conflict-growth nexus originate? Despite many reservations, economic growth constitutes the principal avenue via which sustainable poverty reduction can take place in low income developing countries. Simply redistributing income, without making the cake bigger, only serves to make the already poor more

[1] In many respects, there has been a reversal of fortunes. At the time of independence (1960s) most African nations were ahead of the average East Asian country in real per capita income terms (see Easterly and Levine 1997).

equal. Thus, growth is a necessary condition for poverty reduction in low income countries, and importantly it also serves to stem the seeds of conflict by fostering human development. Lack of growth in many parts of the developing world undermines human security in terms of mankind's freedom from want and fear[2]. More fundamentally, it can promote conflict. There are similarities between conflict prevention and the deep determinants of growth in the long-run because of factors common to both: institutions, inequality, resource rents and so on. Growth failure, or growth biased against certain groups, can produce both greed and grievances, which are necessary (but not sufficient) for the onset of violent internal conflict. Furthermore, in empirical work the most robust predictor of the danger of conflict breaking out is low per capita income, implying low growth (Ross 2004). Thus, an important growth-conflict nexus does exist. Growth, however, is not a panacea for the reduction of conflict risk; certain recent developing country growth experiences, when accompanied by increasing inequalities, especially inequality of opportunity, can also engender conflict. The conflict-growth nexus is, therefore, associated with complex and multi-directional causality: low income makes conflict more likely, conflict negatively impacts growth; highly inegalitarian growth can also produce conflict.

The rest of this work is organized as follows: Long-Run Growth and the Causes of Conflict section briefly looks at the long-term determinants of growth and causes of conflict. Africa's Recent Growth and Conflict Experience section examines the recent experience of Africa in terms of conflict, institutional functioning and economic growth and Conclusion section provides a conclusion along with some policy implications.

2 Long-Run Growth and the Causes of Conflict

2.1 Long-Run Growth

Table 1 shows that growth rates for developing countries as a whole (except in Asia) were greater in the 1960s and 1970s compared to 1980–2000, particularly in Africa, where average growth rates were negative. Since then growth rates for developing countries have picked up, including those for Africa. What are the deep determinants of growth in the long run? We now briefly summarize the long-term determinants of growth, where we have three explanations: geography, culture and institutions[3].

Gallup et al. (1998) have argued that a tropical location, particularly a tropical African location, is disadvantageous to long-term growth. Agricultural productivity is lower in Sub-Saharan Africa because of lack of seasonal variations of the kind that exist in more temperate climates. Secondly, Africa carries a greater disease

[2] The expressions originate in President Franklin Delano Roosevelt's State of the Union address to Congress on 6 January 1941 (see http://www.Fdrlibrary.marist.edu/od4frees.html).

[3] In the short-run the determinants of growth are economic policies to foster capital and skilled labor accumulation.

Table 1 GDP per capita growth rates (annual average %)

Area/country	1960–1970	1970–1980	1980–1990	1990–2000	2000–2010
All developing countries	3.1	3.3	1.2	1.9	5.2
East Asia and Pacific	2.9	4.5	5.9	6.0	8.2
Latin America and Caribbean	2.6	3.4	−0.8	1.7	1.9
Middle East and North Africa	–	–	0.5	1.7	3.1
South Asia	1.8	0.7	3.5	3.2	5.2
Sub-Saharan Africa	2.6	0.8	−1.1	−0.4	3.1

Source: Murshed (2008) for the 1960–2000 period at 1995 constant US$ and author's own estimation (for 2001–2010 at 2010 constant US$) based on data provided by the World Bank's web page on Global Economic Prospect: Regional Outlook, available at: http://www.worldbank.org/content/dam/Worldbank/GEP/GEP2015a/pdfs/GEP15a_web_full.pdf (accessed on 16 April 2015)

burden; especially virulent types of malaria are found there. Finally, there is a feature of the land being locked; a lot of the population in Africa lives at great distances from the sea, making it difficult to transport goods. Cultural explanations for growth hark back to the celebrated study by Weber, who argued that Protestantism instilled values essential for the development of capitalism. In the African case, the saliency of a high degree of ethno-linguistic fragmentation in retarding economic progress was emphasized by Easterly and Levine (1997). The problem, however, with culturalist explanations is that a powerful element of bi-directional causality exists between culture and growth; changes in economic circumstances alter culture.

Acemoglu et al. (2001, 2005) have produced the most influential theoretical basis on the role of institutions in determining long-term growth, which is nowadays widely accepted in economics. They argue that circa 1500, most of the richer countries of the world were located near the equator (including China), whereas nowadays the converse is true, hence institutions rather than geography may be more salient to long-run growth.

Political and economic institutions need to be differentiated. The latter mainly pertain to property rights and contract enforcement, which are associated with the rule of law. Political institutions refer to both formal rules (constitutions or long established conventions), as well as the informal exercise of power. Formal political institutions evolve more slowly as compared to informal power, as is evidenced by the infrequency with which constitutions are altered. Informal political institutions refer to the power of the influential, and are very much related to the distribution of wealth. Political institutions and the distribution of wealth are two factors that jointly determine economic institutions, which in turn determine economic performance or growth, and the future distribution of resources and political institutions. Current economic literature points to several sources of institutional determination, some of which relate to natural resource endowment, others to the historical pattern of colonial settlement, and we also have a role for inequality.

With regard to endowments, certain varieties of natural resources such as oil and minerals are regarded as having the tendency to lead to production and revenue patterns that are concentrated, while revenue flows from other types of resources such as agriculture are more diffused throughout the economy. Mavrotas et al. (2011) present evidence that developing countries which export mineral/fuel type natural resources have tended to have lower growth rates in the post-1970 period[4]. This is known as the resource curse. Natural resource rents can make corruption, predation and rent-seeking a more attractive livelihood option compared to peaceful production in the presence of poor institutions. Equally, a rich mineral type natural resource endowment, where ownership and production is concentrated can engender poor institutions. Malfunctioning institutions may then retard growth.

Acemoglu et al. (2001, 2005) relate poor (or good) institutional determination to patterns of colonialization. They distinguish between two types of colonies. The first group corresponds to parts of the new world settled by European migrants, as in North America and Australasia. The second group refers to tropical developing countries, today's Third World. The idea is that better institutions, especially property rights and the rule of law, were embedded in the first group[5]. In the second category of colonial countries, an extractive pattern of production was set up. As the extractive state is expropriatory and predatory, bad institutions emerge and become entrenched even after independence, and a predatory equilibrium emerges.

Another strand of literature builds on the link between inequality and resource endowment of the mineral-fuel variety. This tends to depress the middle class' share of the income in favor of the elites, as in Latin America and Africa. The idea being that these elites, in turn use their power, identical with the forces of the state, to coerce and extract rents. When different groups compete with each other for these rents, the rent-seeking contest leads to even more perverse and wasteful outcomes than when elites collude, say Mavrotas et al. (2011). Easterly (2007) makes the point that small elite-based societies do not have a stake in the long-term development of the land. Unlike in middle class dominated societies, publicly financed human capital formation and infrastructural development risk neglect, hence depressing growth prospects.

[4] Note, however, that historical evidence regarding natural resource abundance suggests that many of these countries like the United States, Canada, Australia and New Zealand have done well in the past. The latest evidence for the period following the Cold War is also mixed (see Murshed et al. 2015).

[5] The authors argue that the mortality rates amongst Europeans are what determined whether they settled a colony or not.

2.2 Theories of Conflict Onset

In recent years, two phenomena have been used to explain the onset of conflict among rational choice theorists: greed and grievances. The former stems from the influential work of Paul Collier (see Collier and Hoeffler 2004). According to this view, conflict reflects elite competition over valuable natural resource rents, concealed with the fig leaf of collective grievances. Additionally, rebellions need to be financially viable: civil wars supported by natural resource based rents like blood diamonds or oil, or when sympathetic diasporas provide a ready source of finance, are more likely to occur. Above all, there is the assertion that inequality plays no part in starting a civil war. Paul Collier and associates (2003) also emphasize the poverty trap: poverty makes soldiering less unattractive, generally lowering the opportunity cost of war in poor nations.

In turn, conflict serves to perpetuate poverty because of war's destructiveness; a vicious cycle of poverty-conflict-poverty ensues. Fearon and Laitin (2003) assert that ethnic or religious diversity makes little contribution to the risk of a civil war, which is mainly caused by diminished state capacity in the context of poverty. This finding, taken together with Collier's work has a simple intuitive appeal—civil wars occur in poverty stricken, failed states characterized by venal, corrupt and inept regimes, with the dynamics of war sustained by a motivation akin to banditry. Observe that this view has similarities to the resource curse hypothesis described in the previous sub-section. Be that as it may, the greed theory as a cause of conflict is a paradigm in distress, not least because of its empirical shortcomings as an explanation for conflict onset, as opposed to its being a plausible explanation for the length of the war (duration) (see Murshed 2010, chapters 2 and 3 for a literature review). But more importantly, the recent experience of rising inequalities throughout the world has drawn attention to this problem, including its role in producing discontent, mass protests and civil wars. Consequently, the opposing grievance theories of civil war onset may be regarded to have gained ascendancy at present.

There is a long-standing alternative position in political science that relative deprivation (Gurr 1970) and the *grievances* that emanate from it fuel internal violence. Group-based identity is also crucial to intra-state conflict. This is due to the collective action problem (Olson 1965). It is difficult to mobilize large groups to undertake collective action, because of mutual mistrust, monitoring difficulties and the free-rider problem. Ethnic identities, whether based on race, language, religion, tribal affiliations or regional differences, may serve as a more effective amalgam for the purposes of group formation, compared to other forms of more transient differences. The formation of enduring group identities to which individuals subscribe, are therefore central to mobilizing groups, including the machinations of conflict entrepreneurs who organize men to fight each other (see Tilly 1978 and Gurr 2000 on this). Conflict cannot proceed without the presence of palpably perceived group differences, or grievances, which may have historical dimensions. Frances Stewart (2000) has also introduced the notion of horizontal inequality—

inequality between groups—rather than vertical inequality that exists amongst individuals under the assumption of ethnic homogeneity.

Cederman et al. (2013: chapter 2) make a cogent argument that the hitherto cross-national empirical (econometric) literature was largely dismissive of inequality as a causal factor for the onset of a civil war because of the a-theoretical nature of many of these econometric models, and more importantly because of their utilization of vertical inter-personal (household) inequality measures such as the Gini coefficient of household income inequality. A focus on categorical inequalities as suggested by Tilly (1998), or the inequality of opportunities (Roemer 1998), would lead to the deployment of quite different metrics of *group* based inequalities.

Horizontal inequality has many dimensions. Its economic dimension includes discrimination in taxation and the provision of public services, asset inequality, economic mismanagement and local grievances about not obtaining a fair share of resource rents (see Murshed 2010, chapter 3 for a survey). Additionally, there are potentially conflict risk producing political horizontal inequalities; political exclusion of certain groups or ethnicities is crucial in this regard, but not ethnic diversity per se. These factors are summarized in Østby (2013) and in Cederman et al. (2013).

Cederman et al. (2013: chapter 5) use G-Econ data (which is a local level income dataset on geographical grids for 1990), along with a geographically coded political inclusion/exclusion version of their ethnic power relations (EPR) dataset. Using a gap (ratio) measure of income horizontal inequality, and political discrimination variables of inclusion, exclusion and recent displacement they find that both income horizontal inequality, political exclusion and the effect of recent political exclusion significantly affect the risk of the onset of conflict in a global cross-section of regions. The coefficients of the political variables are larger and the interaction between political exclusion and economic horizontal inequality is significant, suggesting that political exclusion and economic disadvantages move together.

Responding to various criticisms, Collier et al. (2009) repackaged Paul Collier's earlier greed argument by devising the concept of feasibility: rebellion occurs wherever feasible, motivation is indeterminate as grievances are in infinite supply awaiting organization. The basic arguments and empirical evidence are, however, much the same as before. Curiously, these arguments about feasibility mirror Tilly's (1978) critique of Gurr's (1970) relative deprivation theory. Tilly had much earlier indicated that grievances are omnipresent; his mobilization for revolt theory requires considerably more than the aforementioned feasibility hypothesis, especially with regard to state capacity. Tilly's mobilization theory is a combination of the various factors considered in the greed and grievance theories summarized earlier, but where an important role is also placed on the psychological framing of the narrative of grievances heralding much later work in behavioral economics (Akerlof and Kranton 2000, for example).

2.3 Synthesis

Greed and grievance bases for conflict may exist simultaneously. Even if a conflict is initially based on a grievance, it can acquire greedy characteristics and vice versa. For example, a civil war originating in demands for land reforms (Colombia, Nepal) can acquire greed based characteristics once the rebels begin to enjoy narcotic rents or tributes from the peasantry. A civil war based on a desire to control lootable revenue rents can also produce grievances as people are killed. In reality the competing greed versus grievance hypotheses may, after all, be complementary explanations for conflict. Grievances and horizontal inequalities may, after all, be better at explaining why conflicts begin, but not necessarily why they persist. Neither the presence of greed nor of grievances alone is therefore sufficient for the outbreak of a violent conflict; this is something which requires an institutional breakdown.

If the forces behind either greed or grievances are to take the form of large-scale violence there must be other factors at work, specifically a weakening of what Addison and Murshed (2006) call the 'social contract' (see also Murshed 2010). This is similar to weak state capacity, and by implication the poor institutional quality arguments made by many. Even if capturable resource rents do constitute a sizeable prize, violent conflict is unlikely to take hold if a country has a framework of widely-agreed rules, formal and informal, that govern the allocation of resources and the peaceful settlement of grievances. Such a viable social contract can be sufficient to restrain conflict, and following its collapse, rebuilding and reconstructing a new social contract will be key to peace building and conflict resolution in Africa as argued in detail in Murshed (2008).

So what factors lead to the breakdown of the social contract within a nation-state? What circumstances create incentives for groups within societies to choose war rather than resolving disputes peacefully? Clearly these seem to occur in failing societies. Yet, the eponymous term 'failed state' may be too vague and unhelpful in this regard. Contemporary African civil wars are more often related to the *break-down* of explicit or implicit mechanisms to share power and resources, rather than the complete *absence* of an agreement to govern these. Among the various factors in question three reasons are highlighted in Murshed (2008). First, there is the imperfect nature of the fiscal redistributive mechanism, whereby a change of political leadership may lead to the system being discriminatory against some groups. Added to this problem is the weak fiscal capacity of the state (its ability to raise revenue), something that civil war diminishes (Chowdhury and Murshed 2014). Secondly, there is economic mismanagement, which causes insufficient or negative growth rates. Finally, the political system can engender conflict, especially during democratic transitions and in imperfect democracies retaining many of the characteristics of autocracies such as a highly imperfect separation of powers between the executive, legislative and judicial functions of the state.

Many of the factors mentioned here also have growth retarding effects. Acemoglu et al. (2005) list the conditions for the development of good institutions that are important in fostering long-term growth. First, when there are constraints on the executive and a balance of power exists between different forces in society, implying a degree of democracy along with a separation of powers. Nowadays, most developing countries, including African states, are imperfect democracies. Although many hold elections frequently, there are few checks on the elected executive. Consequently, the executive tends to exercise unbridled power when in office, and frequently adopts economic policies that only enrich narrow support groups. Some actually set about actively dismantling or undermining checks on their executive powers such as the independence of the judiciary. Secondly, where the enforcement of property rights (necessary for securing investments) are broad based and are not confined to an elite group's interests. Otherwise predation will be common; violence is the easiest means of protecting the vast estates of the wealthy few. In other words, societies with less inequalities and a powerful middle class are more likely to devise superior economic institutions. Finally, fewer capturable 'rents' that can be appropriated by a small group is also a condition for the emergence of more sound institutions. In fact, good institutions can prevent or mitigate the elite capture of resource rents.

3 Africa's Recent Growth and Conflict Experience

Figure 1 illustrates Africa's growth rates from 1980 to 2010. Although the trend line is upwards, there are substantial periods of decline—from 1980 to 2000 on an annual average basis, growth rates were mainly negative for the region as a whole (see also Table 2). Sustained economic recovery took place after 2000.

Since the beginning of the millennium there has also been a decline in both the incidence and intensity (measured by battle related deaths) of civil war in Africa (Table 3). Koubi (2005) points to a 'Phoenix' factor that follows war. This means that there are marked growth spurts in the medium-term following the destructiveness of war. Perhaps this growth is related to improved institutional quality and better governance.

Another factor which helped Africa recover its economic growth was the commodity boom in the first decade of the new millennium. As most African economies are commodity exporters, this trend was certainly helpful. Some progress has been made in transforming African economies to higher value added economic activities in manufacturing, services and the processing of natural resource based products, but concerns remain about the pace of this economic transformation in the light of Rodrik's (2015) findings about premature de-industrialization, which means that countries may be unable to industrialize nowadays, unlike in the past, even at lower levels of per capita income because of both globalization and labor saving technical progress. For example, it means that for Latin American middle income countries that did not industrialize earlier, but relied on natural resource based exports, the possibilities for industrialization are much lower now than a few generations ago due to the rapid pace of industrial

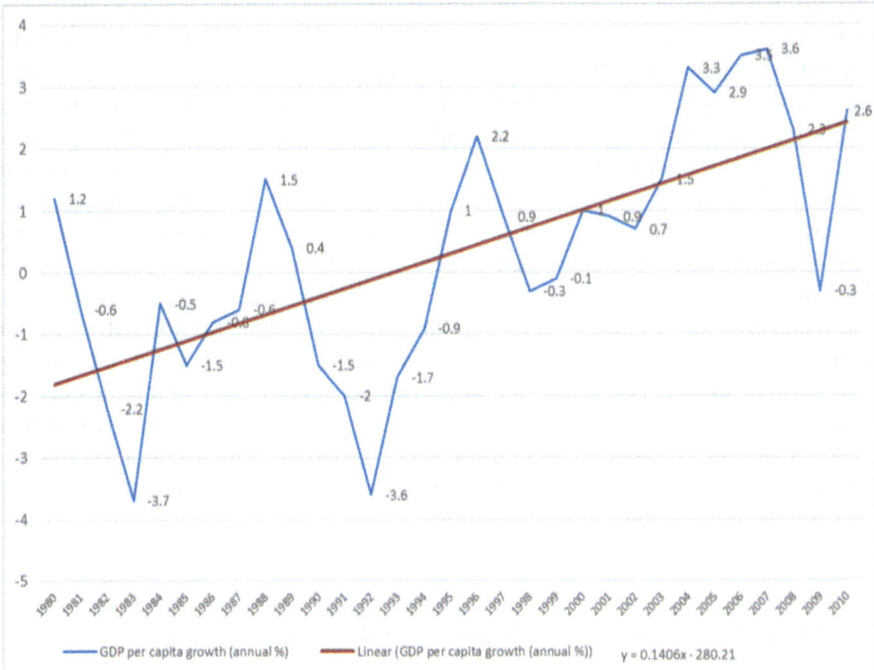

Fig. 1 GDP per capita growth rate in SSA during1980–2010 (annual average %)

relocation, production of components in different locations and labor saving technical progress.

This brings us to the crucial issue of the contribution of institutional quality to economic growth. Table 4 lists growth, institutional quality and conflict indicators for major developing regions for 2000–2010. The second and third columns provide annual average growth of per capita income and the total number of civil war battle deaths for 10 years respectively. The rest of the table is concerned with institutional quality. The polity index is a combined measure of the degree of autocracy or democracy, with autocratic scores ranging from −10 (maximum) to 0 (minimum), and democratic scales varying from 10 (maximum) to 0 (minimum). A combined polity score can indicate an imperfect democracy (or anocracy) with polity scores below 6 or 8 at the upper end and −4 at the lower bound. Most African countries are still imperfect democracies (only Mauritius approaches a perfect democracy). The rest of the table lists the well-known international country risk guide (ICRG) indicators of governance, where a higher score always implies superior quality[6]. The aggregate ICRG index score for Africa is only slightly below that for the total developing country average. As far as the sub-components are concerned, government stability indicates political stability and the ability of the government to carry

[6] ICRG data is missing for many African nations.

Table 2 GDP per capita growth rate (annual %) in Sub-Saharan Africa (1980–2010)

Year	GDP per capita growth (annual %)
1980	1.2
1981	−0.6
1982	−2.2
1983	−3.7
1984	−0.5
1985	−1.5
1986	−0.8
1987	−0.6
1988	1.5
1989	0.4
1990	−1.5
1991	−2.0
1992	−3.6
1993	−1.7
1994	−0.9
1995	1.0
1996	2.2
1997	0.9
1998	−0.3
1999	−0.1
2000	1.0
2001	0.9
2002	0.7
2003	1.5
2004	3.3
2005	2.9
2006	3.5
2007	3.6
2008	2.3
2009	−0.3
2010	2.6

Source: Author's own illustration based on World Bank data; http://data.worldbank.org/data-catalog/africa-development-indicators (accessed 1 April 2015)

out its stated policy goals, and in this respect Africa does well, but not in terms of corruption. With regard to law and order (respect for the law, property rights and contract enforcement) Africa fairs poorly but is ahead of Latin America and the Caribbean. In terms of bureaucratic quality Africa lags behind all other developing regions. Investment profile refers to factors such as the risk of appropriation, and in that category Africa's indicators are better than that for South Asia. It has to be pointed out that governance indicators for Africa have improved in the last two decades, and in that respect institutional constraints to growth in Africa are being eased.

Table 3 Conflict in Sub-Saharan Africa during 1980–2010

Year	Number of countries experiencing conflict	Number of battle deaths
1980	8	NA
1981	9	NA
1982	9	NA
1983	11	NA
1984	10	NA
1985	8	NA
1986	10	NA
1987	11	NA
1988	8	NA
1989	10	34,339
1990	11	64,584
1991	13	25,402
1992	12	6605
1993	9	14,264
1994	11	6866
1995	9	3645
1996	10	6614
1997	13	21,724
1998	14	19,842
1999	12	59,691
2000	13	58,768
2001	14	9992
2002	11	6045
2003	8	5583
2004	6	5490
2005	5	1325
2006	8	3438
2007	9	2919
2008	9	4321
2009	10	4641
2010	8	3708

Source: PRIO-Uppsala Conflict Database, UCDP

Despite a decline in the occurrence of civil wars and improvements in institutional quality, civil wars continue to be a problem in Africa, and continue to retard growth where the intensity of war is high (measured by the number of battle deaths). Table 5 gives growth rates (annual average) for individual countries that have experienced civil war at any time between 2000 and 2010, the intensity of the conflict with their mean polity and ICRG scores. It suggests that nations such as Eritrea experiencing intense conflict still have negative growth rates and a poor polity score. Table 6 gives similar information for African states that did not have civil war on their soil from 2000 to 2010. Even then, some nations had negative

Table 4 Development-conflict-institutional nexus during 2000–2010

Area	GDP per capita (2010 constant US$) annual average growth (%)[a]	Number of battle deaths in civil war[b]	Polity (−10 to +10)[c]	ICRG index (0–100)[d]	Govt. stability (0–12)[d]	Corruption (0–6)[d]	Law and order (0–6)[d]	Bureaucratic quality (0–4)[d]	Investment profile (0–12)[d]
All developing countries	5.2	252,990	3.53	63.93	8.71	2.03	3.07	1.55	7.43
East Asia and Pacific	8.2	19,721	2.89	73.45	9.12	2.70	4.11	2.62	8.93
Europe and Central Asia	4.1	13,056	6.73	75.91	8.60	3.29	4.68	2.85	10.09
Latin America and Caribbean	1.9	8789	6.95	67.99	8.22	2.42	2.73	2.04	7.97
Middle East and North Africa	3.1	26,349	−4.05	72.22	9.63	2.24	4.25	1.97	9.03
South Asia	5.2	99,225	2.01	62.72	8.45	2.20	3.01	2.25	7.07
Sub-Saharan Africa	3.1	107,686	1.85	59.49	8.87	1.94	2.87	1.05	7.13

[a]Author's own estimation based on data provided by World Bank's web page on Global Economic Prospect: Regional Outlook, available at: http://www.worldbank.org/content/dam/Worldbank/GEP/GEP2015a/pdfs/GEP15a_web_full.pdf (accessed on 16 April 2015)
[b]Author's own estimation based on UCDP dataset; available at: http://www.pcr.uu.se/research/ucdp/datasets/ucdp_battle-related_deaths_dataset/ (accessed on 01 April 2015). The best (neither the highest nor the lowest) estimation on number of battle deaths is considered in this calculation
[c]Author's own estimation based on Polity IV dataset available at http://www.systemicpeace.org/inscrdata.html (accessed on 1 December 2014)
[d]Author's own estimation based on aggregate index and political risk component of International Country Risk Guide (ICRG) dataset. The higher value indicates better status and vice versa

Table 5 Civil conflict countries in Sub-Saharan Africa (2000–2010)

Country	Annual average GDP per capita (2005 constant US$) growth rates (%)[a]	Number of battle death	Mean polity score (−10 to +10)	Mean ICRG aggregate score (0–100)
All countries in this category	–	107,686	1.73	57.55
Angola	6.89	2726	−2.14	61.33
Burundi	0.03	5631	4.50	–
Cameroon	0.61	542	−4.00	66.85
Central African Republic	2.31	463	0.36	–
Chad	6.29	4128	−2.00	–
Congo	1.61	167	−4.21	63.57
DR Congo (Zaire)	1.65	3298	3.43	47.61
Eritrea	−2.59	50,057	−6.93	–
Ethiopia	5.14	1276	−1.57	58.74
Guinea	0.36	649	−0.21	56.25
Ivory Coast	−0.48	286	1.43	56.93
Liberia	3.47	2677	4.07	48.08
Mali	2.49	277	6.21	62.30
Mauritania	0.85	190	3.50	–
Niger	0.71	128	4.86	59.38
Nigeria	5.51	533	4.00	59.86
Rwanda	4.67	4290	−3.43	–
Senegal	1.19	119	7.50	63.80
Sierra Leone	2.69	430	5.43	54.36
Somalia	–	7598	0.77	37.79
Sudan	3.03	16,888	−4.67	54.30
Uganda	3.47	5333	−2.07	62.87

[a]The estimation follows annual compound growth rate formulae using data provided by World Development Indicator 2015, available at http://databank.worldbank.org/data/views/reports/metadataview.aspx (accessed on 16 April 2015)

growth rates. But many of these—Gabon, Togo and Zimbabwe for instance—have poor polity scores.

4 Conclusions and Policy Implications

Sustaining growth and minimizing the incidence of conflict in Africa, as elsewhere, is an arduous task. In this connection, traditionally economists and political scientists had different emphases, with the former highlighting the importance of growth

Table 6 Non-civil conflict countries in Sub-Saharan Africa (2000–2010)

Country	Annual average GDP per capita (2005 constant US$) growth rates (%)	Mean polity score (−10 to +10)	Mean ICRG aggregate score (0 to 100)
All countries in this category	–	3.56	62.49
Benin	0.64	6.45	–
Botswana	2.41	8.00	80.93
Burkina Faso	2.61	−0.21	60.74
Cape Verde	4.32	10.00	–
Comoros	–	5.82	–
Gabon	−0.38	−2.73	–
Gambia	0.58	−5.00	–
Ghana	2.89	6.91	–
Guinea-Bissau	0.26	4.45	51.77
Kenya	1.42	6.43	63.09
Lesotho	2.84	7.45	–
Madagascar	−0.36	5.73	–
Mauritius	2.93	10.00	–
Malawi	1.40	5.55	59.17
Namibia	2.98	6.00	77.89
South Africa	1.81	9.00	70.90
Swaziland	1.01	−9.00	–
Tanzania	3.37	−1.00	61.79
Togo	−0.42	−2.71	59.75
Zambia	4.17	5.57	60.18
Zimbabwe	−4.70	−1.93	41.14

[a]The estimation follows annual compound growth rate formulae using data provided by World Development Indicator 2015, available at http://databank.worldbank.org/data/views/reports/metadataview.aspx (accessed on 16 April 2015)

and poverty reduction and the latter stressing institution building and the evolution of sustainable democracies. Yet, as this paper demonstrates, there is an *insepara-bility* between economic policymaking and the political process. Both conflict resolution and sustained economic growth require good institutional functioning in the political and economic spheres. However, it should be pointed out that the relationship between growth and conflict is non-linear; lack of growth and the poverty it produces can engender conflict (most likely civil war), but equally rapid growth accompanied by heightening inequality of opportunities can also lead to conflict (riots, demonstrations and isolated insurgencies) even in the presence of increased state capacity.

With regard to short-term peace building policies, the economic reconstruction that follows conflict needs to be broad-based and pro-poor; otherwise the grievances that produced the civil war may re-emerge. Conflict distorts the economy, making activities with short-term returns such as services more attractive as compared to areas which require long-term investments such as agriculture or manufacturing (see Addison and Murshed 2005). The post-conflict economic recovery may be lop-sided, and based principally on less pro-poor short-term economic activities. This can impose an unfortunate path-dependence on reconstruction and growth. One source of distortion is the sharp increase in transaction costs resulting from war, including the destruction of transport, the planting of land mines and institutional collapse that drive a wedge between producer and consumer prices. Typically, production (especially agriculture) is more vulnerable, leading to a sharper increase in transaction costs as compared to other sectors such as urban-based trade and services. In addition to raising transaction and production costs, conflict raises uncertainty about the future, and therefore the rate of return demanded by potential investors. To avoid this, selective policies of subsidies to the productive sectors may have to be followed.

The centrality of reducing inequalities, particularly durable inequalities between groups (Tilly 1998) based on unequal access to productive assets such as land or education, for lowering conflict risk cannot be overemphasized. The current donor focus seems to be solely on poverty reduction. However, conflict avoidance as well as other developmental considerations such as truly pro-poor growth creates a hugely important link to *inequality*.

In recent years there has been a sharp increase in inequality in all countries in the world, developed and developing. Often, the rise in inequality has been accompanied by impressive economic growth. The rise in inequality of income and wealth has also sharpened and solidified inequalities of opportunities between individuals and groups. Thus, we may say that increased inequalities of income and wealth lead to inequalities of opportunity, as the disadvantaged face even less fair life chances. In this environment, discrimination against groups is likely to either persist or become more acute. Hence, categorical inequalities may become even worse. These are precisely the type of inequalities that sow the seeds of conflict. Therefore, attention has to be paid to redressing the most dangerous form of inequality, categorical inequality, where interventions such as equalizing educational opportunities, take a generation or more to bear fruit.

As far as the political sphere is concerned, refashioning the social contract, especially the institutions of peaceful conflict resolution are important in the immediate post-war period. Chief among these are policies for successful peace treaties and ensuring that they endure by enhancing the commitment of various parties to the agreement. In this connection, policies that are mindful of a fair division between protagonists, sustainable and inclusive power sharing among former belligerents and the adroit management of natural resource rents, where applicable, are crucial (see Murshed 2008). Donors need to be mindful of the fact that the democratic transition has the potential of producing violence in the short-run as people suddenly have greater scope for venting dissatisfaction, especially when this is unaccompanied by sufficient economic progress (Gurr 1970; Hegre et al. 2001). In this connection it has to be remembered that other forms of conflict—mass protests, riots and sectarian conflict—may replace civil war.

External interventions and economic aid in sustaining peace treaties are important in the short-run.

Long-run policies for conflict resolution need to be *sui generis* and endogenous (from within) as Easterly (2006) demonstrates in his study of the overall abject failure of western development assistance in achieving poverty reduction. Solutions that in the long-run are not home grown are doomed. In this regard policies that foster economic diversification and encourage movements away from reliance on a few economic activities (the staple trap) are central to long-term economic progress. In this regard, Rodrik (2006) provides us with a practical check-list for identifying the most binding constraints on growth. These may require removing a distorted economic policy (such as excessive state controls and regulations) or short-term institutional improvement (lessening bureaucratic red tape for example). Economic development in the long-run requires industrialization strategies that make Africa's nascent manufacturing sector internationally competitive, and in this regard there is the danger of premature deindustrialization (Rodrik 2015), unless countervailing policies are pursued. In the political sphere, sustainable (endogenous) democracy usually follows economic growth as part of the process of modernization (Lipset 1960). Democratization without economic growth, like faith without charity, amounts to nothing. Simple multi-party electoral competition in this connection is insufficient. Attention has to be focused on constraining the political elite so that they choose development rather than repression via the separation of powers and an independent judiciary.

References

Acemoglu D, Johnson S, Robinson JA (2001) The colonial origins of comparative development: an empirical investigation. Am Econ Rev 91(5):1369–1401

Acemoglu D, Johnson S, Robinson J (2005) Institutions as the fundamental cause of long-run growth. In: Aghion P, Durlauf S (eds) Handbook of economic growth, vol 1, 1st edn. Elsevier, Amsterdam, pp 385–472

Addison T, Murshed SM (2005) Post-conflict reconstruction in Africa: some analytical issues. In: Collier P, Fosu A (eds) Post conflict reconstruction in Africa. Palgrave, London, pp 3–17

Addison T, Murshed SM (2006) The social contract and violent conflict. In: Yanacopoulos H, Hanlon J (eds) Civil war, civil peace. Currey, Oxford, pp 137–163

Akerlof G, Kranton RE (2000) Economics and identity. Q J Econ 115(3):715–753

Cederman L-E, Gleditsch K-S, Halvard B (2013) Inequality, grievances, and civil war. Cambridge University Press, New York, NY

Chowdhury AR, Murshed SM (2014) Conflict and fiscal capacity. Defence Peace Econ 26, published online 09 Sept 2014. doi:10.1080/10242694.2014.948700

Collier P (2007) The bottom billion: why the poorest countries are failing and what can be done about it. University Press, Oxford

Collier P, Elliot L, Hegre H, Hoeffler A, Reynal-Querol M, Sambanis N (2003) Breaking the conflict trap: civil war and development policy. Oxford University Press, Oxford, The World Bank

Collier P, Hoeffler A (2004) Greed and grievance in civil wars. Oxf Econ Paper 56(4):563–595

Collier P, Hoeffler A, Rohner D (2009) Beyond greed and grievance, feasibility and civil war. Oxf Econ Paper 61(1):1–27
Easterly W (2006) The White man's burden: why the west's efforts to aid the rest have done so much ill and so little good. Penguin, London
Easterly W (2007) Inequality does cause underdevelopment. J Dev Econ 84(2):755–776
Easterly W, Levine R (1997) Africa's growth tragedy: policies and ethnic divisions. Q J Econ 112 (4):1203–1250
Fearon J, Laitin D (2003) Ethnicity, insurgency and civil war. Am Polit Sci Rev 97(1):75–90
Gallup JL, Sachs JD, Mellinger AD (1998) Geography and economic development. NBER Working Paper 6849
Gurr TR (1970) Why men rebel. University Press, Princeton
Gurr TR (2000) Peoples versus states: minorities at risk in the new century. United States Institute for Peace Press, Washington, DC
Hegre H, Ellingsen T, Gates S, Gleditsch NP (2001) Towards a democratic civil peace? democracy, civil change, and civil war 1816–1992. Am Polit Sci Rev 95(1):17–33
Koubi V (2005) War and economic performance. J Peace Res 42(1):67–82
Lipset S (1960) Political man: the social bases of politics. Doubleday, New York, NY
Mavrotas G, Murshed SM, Torres S (2011) Natural resource dependence and economic performance in the 1970–2000 period. Rev Dev Econ 15(1):124–138
Murdoch JC, Sandler T (2004) Civil wars and economic growth: spatial dispersion. Am J Polit Sci 48(1):138–151
Murshed SM (2008) Peacebuilding and conflict resolution strategies in Africa. Background paper for the African Development Bank's annual report
Murshed SM (2010) Explaining civil war: a rational choice approach. Edward Elgar, Cheltenham
Murshed SM, Badiuzzaman M, Habibullah Pulok M (2015) Revisiting the role of the resource curse in shaping institutions and growth. ISS Working Paper no. 605
Myrdal G (1968) Asian drama: an inquiry into the poverty of nations. Allen Lane, London
Olson M (1965) The logic of collective action. Harvard University Press, Cambridge MA
Østby G (2013) Inequality and political violence: a review of the literature. Int Areas Stud Rev 16 (2):206–231
PRIO-Uppsala Conflict Database at www.prio.no/cwp/armedconflict
Rodrik D (2006) Goodbye Washington Consensus, Hello Washington Confusion. A review of the World Bank's economic growth in the 1990s: learning from a decade of reform. J Econ Lit 44 (4):973–989
Rodrik D (2015) Premature deindustrialization. NBER Working Paper no. 20935
Roemer J (1998) Equality of opportunity. Harvard University Press, Cambridge, MA
Ross ML (2004) What do we know about natural resources and civil wars. J Peace Res 41 (3):337–356
Sachs J (2005) The end of poverty: economic possibilities for our time. Penguin, London
Stewart F (2000) Crisis prevention: tackling horizontal inequalities. Oxf Dev Stud 28(3):245–262
Tilly C (1978) From mobilization to revolution. Addison-Wesley, Reading, MA
Tilly C (1998) Durable inequality. University of California Press, Berkeley, CA

MIX
Papier aus verantwortungsvollen Quellen
Paper from responsible sources
FSC® C105338

If you have any concerns about our products,
you can contact us on
ProductSafety@springernature.com

In case Publisher is established outside the EU,
the EU authorized representative is:
Springer Nature Customer Service Center GmbH
Europaplatz 3, 69115 Heidelberg, Germany

Printed by Libri Plureos GmbH
in Hamburg, Germany